Cambridge Elements

Elements in Earth System Governance
edited by
Frank Biermann
Utrecht University
Aarti Gupta
Wageningen University
Michael Mason
London School of Economics and Political Science

COLLABORATIVE ETHNOGRAPHY OF GLOBAL ENVIRONMENTAL GOVERNANCE

Concepts, Methods and Practices

Stefan C. Aykut
Universität Hamburg

Max Braun
Universität Hamburg

Simone Rödder
Universität Hamburg

Shaftesbury Road, Cambridge CB2 8EA, United Kingdom

One Liberty Plaza, 20th Floor, New York, NY 10006, USA

477 Williamstown Road, Port Melbourne, VIC 3207, Australia

314–321, 3rd Floor, Plot 3, Splendor Forum, Jasola District Centre, New Delhi – 110025, India

103 Penang Road, #05–06/07, Visioncrest Commercial, Singapore 238467

Cambridge University Press is part of Cambridge University Press & Assessment, a department of the University of Cambridge.

We share the University's mission to contribute to society through the pursuit of education, learning and research at the highest international levels of excellence.

www.cambridge.org
Information on this title: www.cambridge.org/9781009476041

DOI: 10.1017/9781009387682

© Stefan C. Aykut, Max Braun, and Simone Rödder 2024

This publication is in copyright. Subject to statutory exception and to the provisions of relevant collective licensing agreements, with the exception of the Creative Commons version the link for which is provided below, no reproduction of any part may take place without the written permission of Cambridge University Press & Assessment.

An online version of this work is published at doi.org/10.1017/9781009387682 under a Creative Commons Open Access license CC-BY-NC-ND 4.0 which permits re-use, distribution and reproduction in any medium for non-commercial purposes providing appropriate credit to the original work is given. You may not distribute derivative works without permission. To view a copy of this license, visit https://creativecommons.org/licenses/by-nc-nd/4.0

When citing this work, please include a reference to the DOI 10.1017/9781009387682

First published 2024

A catalogue record for this publication is available from the British Library.

ISBN 978-1-009-47604-1 Hardback
ISBN 978-1-009-38770-5 Paperback
ISSN 2631-7818 (online)
ISSN 2631-780X (print)

Cambridge University Press & Assessment has no responsibility for the persistence or accuracy of URLs for external or third-party internet websites referred to in this publication and does not guarantee that any content on such websites is, or will remain, accurate or appropriate.

Collaborative Ethnography of Global Environmental Governance

Concepts, Methods and Practices

Elements in Earth System Governance

DOI: 10.1017/9781009387682
First published online: May 2024

Stefan C. Aykut
Universität Hamburg

Max Braun
Universität Hamburg

Simone Rödder
Universität Hamburg

Author for correspondence: Stefan C. Aykut, stefan.aykut@uni-hamburg.de

Abstract: Environmental mega-conferences have become the format of choice in environmental governance. Conferences of the Parties under the climate change and biodiversity conventions in particular attract global media attention and an ever-growing number of increasingly diverse actors, including scholars of global environmental politics. They are arenas for interstate negotiation, but also temporary interfaces that constitute and represent world society, and they focalise global struggles over just and sustainable futures. Collaborative event ethnography as a research methodology emerged as a response to these developments. This Element retraces its genealogy, explains its conceptual and methodological foundations and presents insights into its practice. It is meant as an introduction for students, an overview for curious newcomers to the field, and an invitation for experienced researchers wishing to experiment with a new method.

Keywords: global environmental politics, environmental summits, climate governance, global event ethnography, collaborative event ethnography

© Stefan C. Aykut, Max Braun, and Simone Rödder 2024

ISBNs: 9781009476041 (HB), 9781009387705 (PB), 9781009387682 (OC)
ISSNs: 2631-7818 (online), 2631-780X (print)

Contents

1 Introduction   1

2 Ethnography Meets Global Environmental Governance: History and Theory   4

3 Common Conceptual Problems   17

4 Methodological Building Blocks   27

5 Practising Collective Research   37

6 Collaborative Event Ethnography in Action at COP26   50

7 Concluding Remarks   60

References   63

1 Introduction

Collaborative event ethnography (CEE) builds on a series of paradoxes. It is a collective research practice in a field that values individualism and autonomy; it uses a mode of inquiry designed to highlight local situatedness to study world events organised by international bureaucrats in anonymous conference halls; it relies on focused short-term observation missions in highly complex, multicultural environments instead of long-term immersion in a given cultural context. And yet, as we hope to show in this Element, collaborative event ethnography has a unique potential to critically analyse global governance.

Its origins can be traced back to two parallel developments. On the one hand, throughout the 1980s, ethnographers developed new methodological approaches attempting to respond to an increasingly globalised world. Ethnography went global, crafting tools to study transnational elites and circulations. This 'methodological turn' (Hammersley & Atkinson, 2019) also reflected broader developments in political, social and anthropological theory that crystallised around the analysis of networks or the study of governmentalities, and that aimed at uncovering the complex socio-material relations and assemblages that compose world society. On the other hand, the same decades also saw the creation of international organisations and regimes to tackle cross-boundary environmental problems like acid rain, the destruction of the ozone layer, or climate change. Following the 'Earth Summit' in Rio de Janeiro in 1992, environmental mega-conferences became the format of choice in environmental governance. Global climate conferences in particular attracted an ever-growing number of increasingly diverse actors, including scholars of global environmental politics and other social science fields.

CEE as a research methodology emerged at the confluence of these developments. This Element retraces its genealogy, explains its conceptual and methodological foundations and presents insights into its practice. It is meant as an introduction for students, an overview for curious newcomers to the field, and an invitation for experienced researchers wishing to experiment with a new method. A word on terminology: the label *collaborative event ethnography* (coined by Brosius & Campbell in 2010) is today broadly used by scholars who collaborate as a team in order to investigate large but not necessarily 'global' events. However, as we explain in the following chapter, the realm of global governance has emerged as a key site for conducting collaborative event ethnography.

Global environmental conferences bring together a diversity of actors to perform different policy tasks. They are a 'facilitative practice' that holds international regimes together (Lövbrand et al., 2017), as well as temporary interfaces that constitute and represent world society and thus may act as spaces for the co-production of futures (Ibrahim et al., 2024). Observers characterise them as an 'archipelago of meetings' and as 'fuzzy objects' (Dumoulin Kervran, 2021, pp. 82–83). During a climate or biodiversity COP,[1] for instance, no single person can possibly take in all topics and spaces. This makes these events particularly challenging, but also exceptionally interesting sites for ethnographic fieldwork. Of course, these remarks do not apply to all global environmental conferences. Meetings under the umbrella of the Vienna Convention for the Protection of the Ozone Layer and similar treaties are much smaller than climate COPs. Until recently, even biodiversity COPs were much more manageable in terms of participants, side events and accompanying civil society activities. However, climate COPs have come to constitute a focal point of broader environmental debates. To some extent, they provide a model for other global conferences. Throughout the book, we therefore chose to use examples from our fieldwork on climate conferences, especially the COP26 in Glasgow in 2021, to illustrate the practice of ethnographic research on global environmental governance.

Following this introduction, the second section focuses on debates in anthropology and sociology that have informed the emergence of an interest in ethnographic observations of global environmental conferences. This history starts with a diversification of ethnographic approaches. The adoption of some of these new approaches in global governance studies and international relations (IR) expanded the methodological discussion in these communities. Collaborative event ethnography then evolved into an adaptive but recognizable, fully fledged research methodology. Alongside specific practices for data generation and analysis – such as ethnographic observations and qualitative interviews – its research posture aligns with ethnography's traditional self-understanding in terms of a broad and holistic epistemological perspective, grounded in traditions of phenomenology and constructivism.

Section 3 addresses a number of recurring and interconnected conceptual problems and illustrates them using examples from ethnographic fieldwork. UN environmental conferences are global events, and they are also shaped by the culture of a specific host city. They are characterised by mingling and continuous face-to-face interactions, while also constituting mediated 'world events' that synchronise political, science, business and public spheres across the world. They have clear temporal and spatial boundaries, but also constitute condensations of wider developments, and moments in a larger governance process with

---

[1] COP is shorthand for Conference of the Parties, and refers to the regular meetings of the signatories of international treaties such as the United Nations Framework Convention on Climate Change (UNFCCC) and the Convention on Biological Diversity (CBD). In the UN context, COP refers to many conventions and is not restricted to the sphere of environmental governance. In this book, if not indicated otherwise, we use it in reference to climate COPs.

intersessional meetings and preparatory activities. Finally, they represent events with a certain unity, including a well-defined group of accredited attendees, an overarching internal dramaturgy and a corporate design, while also consisting of highly differentiated collections of social spaces with distinct activities and social practices.

In Section 4 we show that, mirroring the complexity of its object of inquiry, collaborative event ethnography is neither an inflexible, monolithic method nor a simple toolbox. It is 'shaped by iterative refinement', through constant exchange among team members (Dumoulin Kervran, 2021, p. 95; see also Corson et al., 2019). In this process, researchers can draw on a set of methodological building blocks, which are each used to a different extent, and with varying accentuations, by different research teams. We identify four key building blocks that adapt traditional ethnography to the specific circumstances and problems associated with studying world events. *Focused ethnography* is used to conduct observations in a short period of time in the field. *Team ethnography* adapts the ethnographic method to larger research collectives, to capture the scale and density of activity of the event. *Digital ethnography* accounts for the fact that events have expanded in physical space and number of attendees, but also into digital spaces. Consequently, it is no longer possible to think about, talk about, and investigate conferences without paying attention to what goes on in the digital sphere. A *dramaturgical perspective* is often used to unearth theatrical elements and performative dimensions of climate conferences and UN summitry.

Section 5 introduces the practice of collaborative event ethnography. Many authors writing on collaborative or team-based approaches in ethnography have emphasised that realising the potential of these approaches is no easy task, and that there is no simple, step-by-step recipe. Collective research stands and falls with the quality of collaborations, and the ability to create a team spirit. It also requires a productive and trusting atmosphere within the collective, in order to enable team members to share experiences and reflect on their emotions while observing. We propose to capture this feature of collaborative event ethnography by identifying a series of collective research practices. Beyond the simple sharing of data and observation notes, these aim at developing capacities for *preparing*, *working*, *thinking*, *experiencing* and *writing together*. Based on a literature review and our own ethnographic experiences, we derive a list of Dos and Don'ts to enable readers to start their own collaborative ethnographic practice.

In Section 6, we illustrate the benefits of collaborative event ethnography with two examples in the form of vignettes, which focus respectively on movement-media interactions at the Glasgow COP conference and the evolution of the Climate Action Zone over several climate COPs. The vignettes were chosen to illustrate two core elements of collaborative event ethnography.

The first is the *principle of many eyes – many minds*, as the presence of many observers and the discussions among them allow researchers to arrive at a richer and more complex picture of a world event. The second is the *principle of repeated observations* within a given global forum, used to identify changes in governance practices that would be invisible to a one-time observer. Both cases also illustrate the capacity of ethnography to capture the emotional and embodied dimensions of global politics, which other approaches struggle to identify and engage with analytically.

In conclusion, we highlight the integrative potential of collaborative event ethnography. Research on global environmental negotiations tends to be separated from studies on transnational governance networks or social movements. Collaborative event ethnography as a methodological approach and collaborative research practice has the unique potential to integrate these fields in observing and writing on environmental world events. It is our hope that this Element inspires its readers to try out and further develop this approach, and in so doing to contribute to a better understanding of the practice of global environmental governance, from its local embeddedness to its overarching social, spatial and temporal dynamics.

2 Ethnography Meets Global Environmental Governance: History and Theory

2.1 Ethnography Goes Global

Historically, ethnography has been associated with qualitative approaches that focus on understanding the meanings and contexts of human behaviour and social interactions. Hallmarks of ethnographic methodology include 'being there' to do first-hand observations (Watson, 1999), 'thick description' of events and situations to add context and meaning (Geertz, 1973), and reflexivity in terms of disclosing subjectivity and working with the positionality of the researcher (Burawoy, 2003). But there has been significant change in what is to be observed, where, in what context, by whom, and how ethnographic work can and should take place, as ethnographic approaches have travelled through disciplinary, geographical and conceptual spaces (Muecke, 1994). These developments have been connected to changes in both the types of social worlds that ethnographers study and changes within those social worlds themselves. At the basic level, ethnography rests on *participant observation*. By immersing themselves in a specific cultural and social context and spending time with people, ethnographers get an 'insider's perspective' on what is happening there. In their seminal introduction to the field, Hammersley and Atkinson (2019, p. 3) write that ethnographers study '[p]eople's actions and accounts [...] in everyday contexts'. Ethnography is a field approach, as opposed to a laboratory approach, or to an approach that rests on studying in the office or the library. Often a variety of data sources is used, ranging from fieldnotes on observations and reflections to interviews and documents, as well as audio and video recordings. Traditionally, data gathering is weakly structured at first, meaning that often the research plan is not fully fixed beforehand, but adjusted progressively, in the course of the research. Data analysis is generally inductive and interpretative, while quantification is not commonly used. Typically, ethnographers look at a small number of cases in depth. As a qualitative social science methodology, ethnography is not a fixed or rigid method, but is most commonly described as an analytic sensibility.

Interestingly, early ethnographic accounts rarely foregrounded methodology or contained explicit discussion of research methods. Rather, good research rested on implicit craftsmanship and embodied skill (Goffman, 1989). Accordingly, in recent decades, the embodied and sensual aspects of fieldwork have come into view as part of ethnography's reflexive repertoire. Ethnographers in both anthropology and sociology have respectively argued for 'thick participation' (Spittler, 2001) and 'observant participation' (Honer & Hitzler, 2015, p. 552), emphasising the need to include an observation of one's own participation in accounts of fieldwork. Moreover, ethnographers writing about their field often refer to an *ethnographic tradition*. This vague outline of features of ethnography, either as something to identify with, continue and develop, or as something to distance oneself from and break free of, comes out of the origins of ethnography as a methodology in early twentieth century anthropology and sociology (Wolcott, 1999, Chapter 5). Defining features of this tradition include that the work is done by a lone researcher, takes place in a specific, locally bounded space, including an extended stay in this *field*, usually of a year or more (see Fine & Hancock, 2017, for an appraisal of the status of this tradition). This approach has at times been tied to a positivist project of capturing the *essence* of people who are very different from the observer, either because of cultural and spatial distance, or because they belong to another social class or cultural milieu. One way of understanding the evolution of ethnography is to explore how conceptions of the field, participant observation and ethnographic writing have changed over time (see Atkinson et al., 2011; Emerson et al., 2011; Markham, 2013; Van Maanen, 2011). We will not explore these developments in depth here, but we do want to highlight three intellectual movements in the history of ethnography that contributed to the emergence of collaborative event ethnography: (1) the call to *study up* towards elites 'at home' in the West; (2) the adaptation of ethnographic work to a globalised world through *multi-sited ethnography*; and (3) the use of *nonlocal ethnography* to bring socio-material constellations and political formations into focus as *apparatuses*.

Studying up

While the origins of ethnographic methodology lie in anthropological research, other social sciences, notably sociology, soon followed suit. Participant observation within rapidly growing and increasingly diverse urban societies was first practiced by members of the Chicago School of sociology in the early decades of the twentieth century. After the Second World War, sociologists like Erving Goffman, Howard Becker, Harold Garfinkel and Anselm Strauss contributed to refining these approaches, giving shape to symbolic interactionism, ethnomethodology and grounded theory as ethnographically inspired sociological approaches to studying Western societies (Fine, 1995). But these early approaches were criticised for their focus on studying groups at the margins of Western society, such as criminals, drug addicts, racial minorities or jazz musicians (Gouldner, 1968). These debates echoed discussions happening in anthropology around the same period. Social anthropologists had begun to criticise the discipline's tradition and practices in colonial and postcolonial societies in Africa (Evens & Handelman, 2006). Members of the Manchester School, for instance, challenged the focus on isolated 'traditional' societies and pointed to the importance of larger political contexts, historical legacies of imperialism, and global interconnections. Within this dynamic of real-world developments and academic debate, Laura Nader's (1972) constructive critique of the ethnographic tradition stands out. Instead of focusing on places far removed from the researcher's experience, located either on the confines of Western imperial geographies or at the inner margins of Western societies, she makes the case for an ethnography of elites. Nader calls for 'studying up' rather than 'down', in order to examine the core institutions of modernity such as police forces, state administrations, insurance companies and commercial practices. In her view, these institutions should be approached with an ethnographic sensibility, combining participant observation with critical examination of the ethnographer's own role. However, Nader also notes the challenges of this endeavour – most notably the problem of field access, for instance in cases where elite communities are closed and unwelcoming to researchers, or reluctant to be the object of investigation.

Sociologists have carried out such ethnographic studies, starting with Garfinkel's (1967) studies on US legal and psychiatric systems, to Becker and colleagues' (1977) work on medical education, Strauss's (1985) writings on hospitals and the medical profession, and Hertz and Imber's inquiries into corporate, professional and political elites (Hertz & Imber, 1995). Influential examples of studying up also include anthropological studies of transnational elites (Marcus, 1983) and so-called *laboratory studies*, which ushered in the formation of the new research field of science and technology studies (STS): Collins & Pinch, (1982), Knorr-Cetina (1981) and Latour & Woolgar (1986). But two decades after Nader's call, anthropologist Hugh Gusterson (1997) noted that ethnographic studies of Western elites and institutions still constituted more of a niche within sociology and anthropology, rather than a broad movement transforming these disciplines. To overcome obstacles to accessing certain fields and popularise studying-up approaches, Gusterson suggested de-emphasising participant observation in favour of 'polymorphous engagement', which for him 'means interacting with informants across a number of dispersed sites, not just in local communities, and sometimes in virtual form', 'collecting data eclectically from a disparate array of sources in many different ways' (Gusterson, 1997, p. 117). Gusterson's re-appraisal of Nader is symptomatic of a turn in sociological and anthropological ethnography in the 1990s towards finding new ways of approaching a globalised and rapidly changing world. Other disciplines also took up Nader's call. Following pioneering work by Fenno (1990) 'watching politicians' in US Congress, ethnographic observation of political elites became a recognised approach in political science, although it continues to sit uneasily within the wider discipline (Schatz, 2009).

Multi-sited Ethnography

Ethnography has traditionally been seen as a methodology to study spatially confined communities and distant cultures. Growing global interconnections, facilitated by waves of economic and political globalisation in the second half of the twentieth century, challenged this focus, which for many constituted the methodological core of ethnography. But globalisation also represented an opportunity to renew ethnographic methods and reaffirm their relevance. As Burawoy points out, 'globalization as the recomposition of time and space – displacement, compression, distanciation, and even dissolution' presents an obvious 'connection to the ethnographer, whose occupation is, after all, to study others in their "space and time"' (Burawoy, 2000, p. 4). *Multi-sited ethnography* emerged as a means to take up the challenge of globalisation and adapt the ethnographic approach to a rapidly changing world (Marcus, 1995). This happened against the backdrop of debates in anthropology about the discipline's historical relation to colonialism, reflexivity, and about textuality, narrative and rhetoric of and in ethnographic knowledge production. George E. Marcus, a key protagonist in these debates (Clifford & Marcus, 1986), points to the need to broaden our understanding of what ethnographic research can and should be, adapting it to what is framed in contemporary discourse as globalisation. Multi-sited ethnography 'importantly [...] arises in response to empirical changes in the world and therefore to transformed locations of cultural production' (Marcus, 1995, p. 97). Contrary to conventional wisdom, Marcus argued that within the very core of the ethnographic tradition, there had already been numerous instances where ethnographers went beyond the local, as they discovered that the communities and cultures they were studying were not as sedentary as they had assumed. Outside of more obvious examples such as migration and diaspora studies, he points to work at the very beginnings of contemporary ethnography, when Malinowski (1922/2004), in a seminal study, accompanied the Trobriand Island 'Argonauts' on their cyclical voyages. Building on this tradition, Marcus argues that following the people, the thing, the metaphor, the conflict, etc., across multiple sites within a connected world is an imperative for ethnography. Crucially, however, this does not necessarily mean constant mobility on the side of the ethnographer. Multi-sited ethnography is not the polar opposite of 'strategically situated (single-site) ethnography' (Marcus, 1995, p. 110). Instead, it involves identifying nodal points in networks, 'system-awareness in the everyday consciousness and actions of subjects' lives' (Marcus, 1995, p. 111), and overcoming distinctions between the local site and the global system, or the ethnographer's 'field' and its political and social 'context'.

Nonlocal Ethnography

Marcus' proposal to adapt ethnography to a world of increasingly complex sociospatial constellations has been exceptionally influential (Coleman & von Hellerman, 2011; Falzon, 2009; Hannerz, 2003). Gregory Feldman (2011) takes some of its core ideas one step further in his plea for a nonlocal ethnography. A scholar of Foucauldian governmentality studies, Feldman argues that it is not enough to follow the movements of actors, ideas or artefacts across different physical sites. Some of the phenomena worthy of ethnographic attention, he points out, are held together not by direct material connections between different localities, but by less tangible ties: symbolic, organisational or social relations that are much harder to observe or follow. To characterise these ties, Feldman uses the notion of the *apparatus*, a term widely used to translate Foucault's concept of *dispositif* (Rabinow, 2003, pp. 49–55). The term denotes 'a thoroughly heterogeneous ensemble consisting of discourses, institutions, architectural forms, regulatory decisions, laws, administrative measures, scientific statements, philosophical, moral and philanthropic propositions [...].

The apparatus itself is the system of relations that can be established between these elements' (Foucault, 1980, p. 194). Feldman proposes the use of the concept of apparatus to grasp 'how unconnected actors are nevertheless related in social constellations' (2011, p. 380). Nonlocal ethnography then becomes a means to study decentralised or polycentric forms of governance by combining ethnographic methods with analyses of documents and discourses. Feldman operationalises the method and demonstrates its usefulness by examining the EU border and migration regime. In this case, as in other governance fields or complex social constellations, relying solely on participant observation is difficult. The involvement of multiple and diverse actors requires researchers to rely on different types of data and combine different methods of analysis.

2.2 The Rise of Global Environmental Summitry

Global environmental governance can be seen as a prime example of an apparatus in Feldman's terms (Wilshusen, 2019). Composed of a multitude of legal frameworks, agreements, actors at different scales, conventions and organisations, touching on all spheres of political, economic, and cultural life, it is notoriously difficult to grasp. Its origins can be traced back to the creation of international regimes for environmental protection in the 1970s and 1980s, starting with treaties on the protection of endangered species (1973), long-range transboundary air pollution (1979) and the protection of the ozone layer (1985). From the outset, the adoption of environmental treaties and the creation of international organisations has been closely linked to the format of large international conferences. Key early examples include the UN Conference on the Human Environment in Stockholm in 1972 and the UN Conference on Environment and Development (or 'Earth Summit') in Rio de Janeiro 20 years later. These meetings have progressively taken centre stage in negotiations over the shape of global environmental governance, today most prominently in the form of the climate and biodiversity COPs (which were adopted at the Rio Conference). They became the 'place to be' for both practitioners and scholars from different disciplinary backgrounds who are interested in some aspect of global environmental governance.

The increasing importance of international conferences in post-WWII governance has been called 'summitry' (Dunn, 1996). More broadly, it has been argued that historically, the development, formalisation, refinement and general proliferation of the social form of the meeting is a central characteristic of many aspects of modern organisational and political life (Brown et al., 2017; van Vree, 2011). Transnational 'mega-events' themselves date back to the nineteenth-century World's Fairs, which in turn constitute a revival of the ancient tradition of Olympic Games. Addressed to a multinational audience, such events have played a key role in the emergent construction of a 'world society' (Meyer, 2010) or 'world polity' (Dumoulin Kervran, 2021, p. 81) through logistical and communicational efforts aimed at constructing 'greatness' (Dumoulin Kervran, 2015). Roche identifies them as core elements of modernity, in which transnational elites celebrate 'civilisation' and 'progress' while simultaneously relying heavily on ritualised elements, dramaturgy and spectacle before a global public (Roche, 2000, p. 89; see also Merry, 2006, pp. 981–2). Similarly, UN environmental conferences have been described as circus (Aykut, 2017), theatre (Death, 2011), or ritual (Little, 1995). As highly mediatised 'world events', they 'provide an arena where states are kept under pressure through the self-binding nature of publicly stated declarations (Heintz, 2014), through informal sanctions of blame games, through the competition for 'soft' global goods such as legitimacy, attention and prestige (cf. Werron, 2015) and through the mobilisation of scientific evidence (Schenuit, 2023, p. 171)' (Ibrahim et al., 2024, p. 7).

At the beginning of the millennium, Seyfang (2003) surveyed the history and functions of mega-conferences. He noted that it is not the all-encompassing conferences of the 1990s, but rather the 'single-issue' (p. 225) formats that came out of Rio 1992 – most notably the United Nations Framework Convention on Climate Change (UNFCCC) and the Convention on Biological Diversity (CBD) and their respective Conferences of the Parties (COPs) – that have proven most consequential in shaping global governance. What makes these conferences stand out is the diversity of actors who attend and the different kinds of activities they involve. This is particularly true in the global climate arena, where changes in the governance architecture following the breakdown of climate negotiations at COP15 in Copenhagen in 2009 actually increased the importance of UN conferences. The adoption of the Paris Agreement at COP21 in Paris in 2015 was widely celebrated as a sea change in global environmental governance. It set up a framework that replaces the previous emphasis on top-down targets with a system of self-determined pledges, regular reporting cycles and public review mechanisms for states and private actors. This foregrounds 'summitry' and theatrical aspects of global climate governance (Death, 2011), as the success of the Paris approach rests not only on institutional frameworks and procedures but also on soft coordination through dramaturgical and performative strategies (Aykut, Schenuit, et al., 2022). In climate conferences, side events have been growing in importance over recent years. To be sure, more public attention does not automatically translate into greater public scrutiny or accountability. For example, rather than providing occasions to 'blame and shame' governments and firms for their lack of ambition, side events at

COP25 in Madrid have been used by diplomats and business representatives to 'claim and shine', emphasising punctual successes and isolated 'best practices' in front of a global audience (Aykut, Schenuit, et al., 2022).

Despite these and similar critiques, climate COPs, with their massive attendance, multiplicity of stakeholders and large civil society spaces, have come to provide a model for other global environmental conferences. In this model, the traditional goal of UN conferences – negotiating international agreements – is sidelined to some extent, while accompanying activities take centre stage. Ever more numerous and inclusive meetings of an increasing number of NGO representatives and other non-state actors that surround environmental summits have become a major attraction in their own right (Lövbrand et al., 2017). Participants view these events as chances to network and facilitate contacts. Civil society spaces have cropped up around the official negotiations, in addition to the (more or less) grassroots initiatives that spring up around mega-conferences. Besides meetings of government officials and diplomats, business, and activists, environmental summitry is also animated by UN administration and the interplay of administrators and bureaucrats with other actors in the 'global' arena (Saerbeck et al., 2020; Well et al., 2020).

The distinctive contribution of ethnographic studies of environmental governance with respect to studies using other approaches is their attention to empirical nuance (Büscher, 2014), and their way of taking these sites seriously as objects of investigation in their own right. Transnational mega-events in earth system governance have been met with criticisms since their very beginnings (Blühdorn, 2007). But collaborative event ethnography as an academic project brackets the question of the efficacy of these conferences, at least temporarily and for analytical purposes, examining them not only as a locus for negotiations and agreement-making, but also as arenas for spectacle, ritual, and social interaction. Contributing to the opening of this investigative space, Death (2011) positions 'summit theatre' as a phenomenon in its own right – one that is ripe for empirical investigation. In his view, 'the theatrical summit becomes a tool of government in this regard, an attempt to inspire and conduct the self-optimisation of the watching global audience' (p. 7), and an 'exemplary form of government [that] has particular effects in terms of the construction of sustainable subjects and the disciplining of participation and engagement. The focus upon inspirational examples prioritises highly visible and hierarchically situated actors' (p. 15). This feature of environmental mega-conferences makes them highly relevant for researchers who are interested in studying up. Although early conferences did not explicitly give accreditation to researchers, the progressive inclusion of civil society observers provided access for ethnographic participant observation.

2.3 The Main Approaches to Collaborative Event Ethnography

Ethnographic observations of global environmental conferences have formed part of a larger 'ethnographic turn' in the discipline of IR, which has been based on importing data collection methods, writing styles and theoretical sensibilities (Vrasti, 2008), including practice-focused research, autoethnography and multi-sited studies, from other disciplines (Montsion, 2018). The ethnographic turn was driven by IR scholars in search of methods to investigate the 'microphysics of power' (Neumann, 2007, p. 192) and track transnational phenomena across multiple sites (MacKay & Levin, 2015; Neumann, 2007). Conversely, anthropologists had become increasingly interested in the international as a field of research, because their local terrains were affected by the activities of international organisations, or because their informants or objects of research became entangled in international arenas (Bellier, 2012; Müller, 2012; Müller & Cloiseau, 2015). Ethnography also helped to foster interpretative orientations to knowledge production beyond causal models (Jackson, 2008), and opened up new avenues for the development and expression of critical and emancipatory perspectives (Vrasti, 2008).

A common starting point of ethnographic approaches to multilateral negotiations is a view of these processes as far more complex than is allowed for by the simplified models of 'negotiation games' between rational actors that feature prominently in dominant accounts of global governance in economics and political science (Müller, 2013). Instead of negotiations between governments, and governance outcomes such as treaties or decisions, they expand the focus to a multiplicity of actors, embodied practice and emotional experience. This approach strongly resonates with process-oriented perspectives that have been on the rise in IR (Bueger & Gadinger, 2014). Scholars working in this line have turned to analysing governance as social practice that unfolds in time and space, functions according to specific logics and norms, produces meaning, and creates artefacts that circulate and are taken up, but also reinterpreted and readapted, in local contexts. Such accounts shift attention towards the question of how international negotiations and global summits make global problems 'governable' in multilateral settings (Müller, 2012). Seen from this perspective governance appears to be not so much about *solving* problems as about *managing* them (Hoppe, 2011) by framing new issues in the terms of dominant institutional orders, cultural norms and organisational routines (Gusfield, 1984; Lascoumes, 1994). It also resonates with a growing attention to emotions in IR. Of course, practices of international affairs and world politics are rich in individual and collective emotions, such as anger, anxiety and hatred, but also pride, joy and hope (Ariffin et al., 2016; Clément & Sangar, 2018; Koschut, 2020).

Scholars attempting to theorise emotions in world politics (Hutchison & Bleiker, 2014) have found patterns of 'institutionalized passions' (Crawford, 2014) and 'feeling structures' (Koschut, 2020), which reflect underlying normative and power structures. But approaches in IR centring on statistics, document analysis or interviews can only indirectly capture this layer of international affairs, through emotional discourse analysis (Koschut, 2018) for example. Ethnographers of global conferences can go further by directly observing and experiencing the emotional dimension of such conferences, and by reflecting on their own feelings around such events in the context of accelerating climate and environmental destruction.

These debates and lines of thought have been taken up by different groups of scholars studying global environmental conferences. A review of the genealogy of collaborative event ethnography (Dumoulin Kervran, 2021) identifies two broad lines of scholarship, in which researchers – often from multiple disciplinary backgrounds – go into the field as a group, collaborate in data gathering and analysis, and publish together while also pursuing their own research and publication interests.

The 'Duke School' of Collaborative Event Ethnography

A first group formed in the early 2000s at Duke University in the U.S. State of North Carolina. It is this group that coined the term collaborative event ethnography. The group includes researchers from geography, political science, anthropology and other fields with a shared interest in the global environmental politics of conservation. The Duke University group has focused mainly on conservation and biodiversity, performing fieldwork at various relevant conferences. According to Brosius and Campbell (2010, p. 248), the idea behind the decision to take a collaborative approach to the study of large environmental meetings was to mirror and mimic the strategy of national delegations and larger NGO observer groups at such meetings: arriving at the event in teams, visiting various events, exchanging ideas and comparing notes about their impressions at informal debriefing sessions. The group formed at the occasion of fieldwork at the 4th World Conservation Congress in Barcelona in 2008. Its approach and experiences are related in a special issue of *Conservation & Society* (Brosius & Campbell, 2010) and in a subsequent special issue of *Global Environmental Politics* dedicated to reports on the 10th biodiversity COP (Campbell, Corson, et al., 2014; Corson et al., 2014). They have also conducted fieldwork at the Rio+20 conference in 2012, the 2014 World Parks Congress, and the 2016 World Conservation Congress (Corson et al., 2019). The starting point of the group's approach is the intuition that studying meetings is a useful way to understand global environmental governance beyond its written conventions (Campbell, Corson, et al., 2014). Their ethnographic approach was an innovation in this context of governance processes that involve entanglements of private and public actors, NGOs, transnational networks and international organisations. Underlying these processes, we find a shift in global environmental politics towards 'advanced liberal governmentality', in which regulatory functions are stripped away from states and moved to international arenas, which become producers of soft rules and guidelines (Death, 2011, p. 2). Collaborative event ethnography is positioned as a methodological answer to these transformations. Its main assets are empirical nuance (Büscher, 2014) and a focus on contingency, historicity and the multiplicity of actors populating complex multi-level governance arrangements (Gray, 2010; Hitchner, 2010). Ethnographic observations at global environmental conferences serve to bridge the gap between paying attention to 'structural forces' and focusing on contingency as 'particular actors use political space in pursuit of outcomes at certain moments in time' (Corson et al., 2014, p. 27). They allow the 'black box' of decision-making at international fora to be opened up, shedding light on the circulation of ideas, the formation of alliances and the making of compromises (Duffy, 2014, p. 127).

This group of scholars has brought its empirical focus to bear on a wide range of contexts, such as discussions around biofuels (Maclin & Bello, 2010; Scott et al., 2014), relations between conservation and climate change concerns (Hagerman et al., 2010), ocean governance (Gray et al., 2014; Silver et al., 2015), development concerns in conservation policy (Peña, 2010), and the role of Indigenous groups and Indigenous knowledge in conservation discourses (Campbell, Hagerman, et al., 2014; Doolittle, 2010; Monfreda, 2010). Questions on the negotiation of the event ethnographer's role as social scientist vis-à-vis the professional community are central in each case (Welch-Devine & Campbell, 2010). This work also draws systematically on concepts from STS such as boundary work, boundary objects, and the co-production of scientific and political orders (Gray et al., 2014; MacDonald & Corson, 2012).

Over fifty researchers from political ecology, geography, anthropology and political science have been involved over time in the Duke project of collaborative event ethnography. Their research has yielded a vast number of case studies and theoretical contributions. In a survey article that reflects on common threads and characteristic challenges running through this body of work, Gray et al. (2019) highlight long-term engagement, collective research and writing, and mentoring opportunities as key benefits of the approach. The challenges they identify include the difficulty of finding common publication outlets while accommodating different publication strategies, as well as the need to overcome disciplinary and personal differences. In another piece reflecting on the approach, Corson et al. (2019) identify institutional ethnography and multi-sited ethnography as common methodological inspirations. They further emphasise the importance of a shared conceptual framework that loosely binds together the different interests and theoretical sensitivities that researchers bring to collaborative work. As an example of such a concept, the authors give the notion of assemblage, which, they argue, serves the aim of composing a bigger picture of conservation governance out of individual case studies.

A French Tradition of Global Event Ethnography

The second group identified by Dumoulin Kevran (2021) is a more heterogenous set of scholars, largely based in France and including social movement researchers and scholars of global environmental governance. In 2007, Marie-Emmanuelle Pommerolle and Johanna Siméant led a group of over twenty-five researchers to study the Seventh World Social Forum in Nairobi (Pommerolle & Siméant, 2011). The team used a common observation guide and relied on binational French and Kenyan observation duos to sensitise researchers to cultural nuance and discussions among international activists from the Global North and Global South. While this study continued to rely heavily on quantitative methods, the next iteration of the method at the Tenth World Social Forum in Dakar in 2011 more explicitly employed qualitative ethnographic methods (Siméant et al., 2015b). Conceptualising World Social Forums as 'knots' or 'coral reefs' (Tarrow, 2011, p. 246) of transnational activism, where social movement actors coalesce, the authors conducted 'a methodological reflection on how to conduct a sociological survey in an international context', and how to think about 'the division of activist labor in this specific context' (Siméant et al., 2015a, p. 11). Quantitative methods such as large-scale surveys and multiple correspondence analysis were complemented by qualitative approaches aimed at examining performative aspects of transnational activist summits, symbolic processes of exclusion and inclusion, and the distribution of social roles among activists (Baillot et al., 2015; Herrera et al., 2015). The authors stress the importance of constant exchanges among the team as a means to overcome challenges associated with short-term observations, and foreground the collective research process as a genuinely productive research method (Siméant et al., 2015a, p. 13).

In parallel to these social movement researchers, a Paris-based team of climate governance and science studies scholars conducted observations at climate conferences COP14 in Poznan in 2008 and COP15 in Copenhagen in 2009. The stated objective was to render a 'thick description' of the conference 'to convince the reader that climate arenas offer an interesting window for understanding how the contemporary world operates in the face of this [i.e., the climate] problem' (Dahan et al., 2009, p. 2, our translation). This perspective enriched analyses of climate negotiations themselves with a focus on the activities of NGOs, business actors and experts, and by tracing the co-production of scientific and political orders in climate arenas (Aykut & Dahan, 2011; Dahan-Dalmedico, 2008). In a metaphor borrowed from the Avignon theatre festival, the authors describe the social and spatial configuration of UN climate conferences as composed of an inner circle of negotiations ('in'), an official accompanying program me of country pavilions and side events ('off') and an outer circle of events organised in hotels and event locations near the convention centre ('off du off') (Dahan et al., 2009, pp. 6–7; Dahan et al., 2010). To capture the increasing importance of activist happenings and social movements, this outer circle has been renamed 'the fringe' in a more recent study on the Glasgow climate conference COP26 (Aykut, Pavenstädt, et al., 2022).

On the occasion of COP21 in Paris in 2015, the team joined forces with a third group of researchers, also mostly based in France, who were working on biodiversity and agriculture governance. This group had used ethnographic methods to observe and analyse the involvement of Indigenous peoples, NGOs, trade unions and business groups at the Rio+20 Summit on Sustainable Development in 2012 (Foyer, 2015c). They viewed the Rio+20 conference as a 'moment of crystallisation' and an 'entry point for examining a series of global issues, including the governability of the planet, the links between capitalism and the environment and the constitution of a global civil society' (Foyer, 2015a, p. 4, our translation). The group built its work together around a view of the Rio summit as a testing ground for ecological modernisation. The encounter between the Rio+20 and climate governance teams at climate COP21 demonstrated another virtue of collaborative work: the potential for new insights to emerge out of encounters between different conceptual and empirical perspectives. While members of the Rio+20 team 'specialising in areas not directly related to climate change observed the climate regime's growing influence on their objects of study, the second group witnessed how the climate regime increasingly took on new issues' (Foyer et al., 2017, p. 12). This led them to formulate the hypothesis that the Paris conference formed part of a broader two-fold trend of 'climatisation of the world' and 'globalisation of the climate' (Aykut et al., 2017). This broad hypothesis accommodated a diversity of methodological and conceptual approaches, and different empirical foci. For instance, the analysis of climate negotiations showed how the inclusion of new issues led to an extension of climate governance networks (Aykut, 2017). And analyses of side events uncovered a process of symbolic exchange between Indigenous leaders and UN bureaucrats, resulting in a 're-enchantment' of climate conferences by Indigenous cosmologies (Foyer & Dumoulin Kervran, 2017). Researchers from this group also examined economic actors' attempts to form a united 'business voice' at COP21 (Benabou et al., 2017), social movements' struggles to construct alternative globalities within a decentralised climate governance landscape (de Moor, 2021), and the role of philanthropic foundations in shaping the discursive context of climate governance (Morena, 2021).

In 2015, an international methodological seminar on mega-events in Paris saw a consolidation of global event ethnography as an emerging research field, facilitating connections and the formation of a research network tied together by a shared interest in the ethnographic study of mega-events (Dumoulin Kervran, 2021). This loose network is characterised by very different foci and research interests, but united by a shared methodological interest in observing the practice of global environmental governance and its diversity of actors through an ethnographic lens.

3 Common Conceptual Problems

3.1 Situated Globality

Environmental problems often have transboundary aspects, and some have come to be viewed as quintessentially global in nature. Anthropogenic climate change, for instance, has been scientifically framed as a global problem by way of global climate models and by the assessment reports of the Intergovernmental Panel on Climate Change (IPCC). Accordingly, a common assumption that has driven the creation of an international governance regime for climate change under the umbrella of the United Nations holds that any political response to it must also be global (Aykut, 2020). But what is 'the global'? 'Globality' stands in a peculiar tension with the local; as Michael Burawoy notes, 'what we understand to be "global" is itself constituted within the local; it emanates from very specific agencies, institutions and organizations whose processes can be observed firsthand' (Burawoy, 2001, p. 150). This characterisation fits with global environmental conferences, and especially climate COPs, which are exemplary sites for the 'making of globality' (Foyer et al., 2017, p. 2). Akin to a travelling circus, they are built from scratch in a new city every year, then unmade only to be reassembled in a new location the following year. As Foyer et al. (2017) note, the UN globality enacted at climate conferences shares some characteristics with Augé's (1995) 'non-places' of globalisation, whose extension relies in large part on the global spread of functional localities devoid of local specificity and meaning. 'Like shopping malls, successive climate conferences reproduce the same delocalised spatial and organisational arrangements. Like airports, they are detached from their surroundings through physical barriers and security checks. Like hotels, they are spaces of transit that are essentially defined by the functions that they accomplish. The venues of climate conferences are the infrastructure of what is commonly referred to as "global governance'" (Foyer et al., 2017, p. 3). However, COPs are also part of a history of meetings with a dramaturgical arc made up of disappointments, transitions and historic breakthroughs. They attract a similar group of people every year, creating repeated interactions and forming part of personal biographies. Although they function according to UN rules, norms and procedures, they are also situated, imprinted by local political agendas and cultural specificities of the host city and country. They are thus sites where the global is produced and enacted, and where local and national concerns are brought into the context of UN governance agendas. One way this has been captured is through the notion of *climatisation*, which denotes a specific form of synchronisation of discourses, agendas and practices that occurs at COPs, but also extends beyond the conference spaces (Aykut & Maertens, 2021). The global nonetheless takes on a specific shape and appearance at each UN summit, climate COP or biodiversity meeting. Enacted at a specific site, 'global' UN culture is coloured by the 'local' host city and country (Aubertin, 2015; Dumoulin Kervran, 2015). The result of this superimposition can be illustrated using observations made at the climate COP26.

Turning Glasgow into a Stage for Global Climate Governance

The conference temporarily transformed its host city of Glasgow (Aykut, Pavenstädt, et al., 2022). Not only were the UN insignia, flags and COP logos visible throughout the city's streets and squares; the whole city was set up as a stage for global climate governance. The area surrounding the conference centre had been intensively cleaned and tidied up. Advertisements throughout the city, from street posters and billboards to the windows of individual stores and shopping malls, all related to the COP and climate themes. This included a series of messages of urgency in public transport and billboards, as well as ScotRail's omnipresent self-depiction as a 'Net Zero Hero' (Figure 1a,b). Those registered at COP26 were given transport cards that allowed free travel throughout Glasgow's public transport network.

As with other COPs, all climate references disappeared from Glasgow overnight as soon as the negotiations ended, making space for the usual Christmas advertisements. The temporary superposition of global and local agendas produced tensions, but residents were also able to use it to their advantage. On the one hand, strict security measures made life difficult for locals. The conference area was located in lively Finnieston, an area not far from the University of Glasgow and a popular

Figure 1 ScotRail's self-depiction as a Net Zero Hero during COP26 in Glasgow (a, b)

residential area for students. As access to the Blue Zone was highly controlled, conference activity and security measures spilled out into the neighbourhood, blocking some residents' access to their homes and forcing them to take long detours around police barriers when protest occasioned increased police presence. Residents' attention was further drawn to the presence of a world event in their neighbourhood by an exclusive 'world leader's dinner' held at the beginning of the conference in nearby Kelvingrove Art Gallery. Again, tight security measures entailed closing down residential streets in the vicinity, forcing residents to cross Kelvingrove Park at night, with no lighting provided. On the other hand, ScotRail staff seized the window of opportunity before the conference and threatened strike action during COP26. As some locals (and hotel chains) were able to jack up the prices they charged for rented accommodation exorbitantly, this was especially concerning for delegates, observers and activists who had decided to stay in less overbooked Edinburgh, an hour's train ride away from the conference. Action affecting this commuting line would have been disastrous for the conference organisers, and at the last minute, a deal for significantly higher pay was concluded between ScotRail and the RMT union. Another strike hit Glasgow's waste disposal services, as the GMB union launched industrial action during the summit, forcing the city to hire many private waste disposal companies to keep the city's bins from overflowing while the world event was in progress.

3.2 Co-presence in a Media Age

As argued earlier, global environmental conferences are about more than negotiations (Little, 1995). Regular interactions and 'meetingness' (Boden & Molotch, 1994; Urry, 2003) give rise to a unique milieu of social actors, a governance community with its own culture, rules, rites and practices.

The co-presence of otherwise geographically dispersed communities is a key motivator for conference attendance more generally – there is no real replacement for bringing people together in one place for a certain amount of time (Collins et al., 2023; González-Santos & Dimond, 2015). This makes UN conferences ideal field sites for ethnographic observation. 'Being there' is inescapable for those seeking to understand 'what's going on'. Networking and informal contacts are important features of such meetings, which have come to encompass a variety of spaces that resemble trade fairs, business exhibitions or popular happenings (Obergassel et al., 2022). Meetingness is also captured in the importance of catering, as lunches, dinners and cocktails are important facilitators for informal networking. But this interaction infrastructure is highly mediated. At the conference site, digital screens display the day's tightly packed negotiation agenda and side event schedule, and indicate the location of each event (Dahan et al., 2009). Moreover, as 'world events', COPs are watched, analysed, interrogated and commented on by global media (Figure 2a,b,c); they attract the attention of different actors across the globe, and align agendas and discourses across distinct social worlds (Death, 2011; Ibrahim et al., 2024).

Each UN conference has its own milieu of international diplomats, business representatives, reporters, bureaucrats, scientists and activists, grouped under the broad labels of 'parties', 'media' and 'observers'. Those who have been socialised at climate COPs, for instance, form 'a small world of sort of "COP-junkies", a social world of global climate change experts [. . .] [with] a strong professional *illusio* [. . .]. They made COPs and COPs made them' (Dumoulin Kervran, 2021, p. 91). Although heterogeneous, this milieu has become deeply familiar with the

Figure 2 The slogan 'The World is Looking to You COP26' in Glasgow's streets (a, b) and subway stations (c)

topics, language, actor constellations and procedures of global climate conferences. This capacity to bring together social worlds and create a space of intense interactional activity is a core feature of global environmental conferences. Their potential to 'synchronise' agendas and practices has also been identified as a key factor for their success (Laux, 2017). In his account of COP21 in Paris, Laux points out that the conference succeeded in synchronising the activities and interests of key actors from political, scientific, corporate and media fields, who otherwise have conflicting interests and operate under asynchronous temporalities. This temporary alignment made possible the adoption of the Paris Agreement, which in turn aims at further aligning sectoral interests and timescales through decarbonisation pathways compatible with the agreement's temperature goals. The adoption of the 'aspirational' 1.5°C goal provides an interesting example of tensions that can arise while synchronising typically asynchronous domains, in this case politics and science. Guillemot (2017, p. 52) argues that the adoption of the 'policy-driven' 1.5°C target, as opposed to the more 'science-driven' 2.0°C target, was the result of a decades-long process of political negotiations and science–policy interactions, culminating in the Paris approach's emphasis on bottom-up pledges, long-term goals and scientific assessments, instead of binding emissions reduction commitments and timetables for implementation.

Cycles of Hope, Expectations and Media Attention

But despite these aspects of meetingness, co-presence and synchronicity, media are also key actors in climate governance. Reporters and cameras present at conferences shape the ways in which attendees act and present themselves, creating a peculiar dramaturgy for public events that is aimed at 'producing good pictures'. International media also shape public discourse and contribute to the transnational visibility of climate COPs and UN summits. Media coverage of climate summits has been shown to constitute a key driver of public interest in climate change (Schäfer et al., 2014). Death (2011) presents a portrait of the media's specific COP attention cycle. During the first days, which usually include the so-called 'world leader's summit', media reports conjure the 'last chance to save the world' and contribute to a build-up of expectations of the 'most important COP ever'. After the conference opening, hopes and expectations are gradually dampened, with criticism becoming louder as the first week goes on. Demonstrations and civil society events midway through a conference, such as the Global Day of Action at climate COPs, regularly feature opposing views among participants, with negotiators expressing hope and praising advances in the negotiations, and activists denying any sign of progress or dismissing negotiations as inconsequential talking shops. Subsequently, as the end of the conference approaches, a new form of expectation management sets in. Possible landing grounds for an agreement are identified, and all attention is focused on negotiators' last-minute struggles to come to an agreement, which is often only achieved in 'extra time', after repeated deadline extensions. Finally, a typically sobering review of what happened makes the news, including assessments that not only cite some progress, but also stress the expectations that were not met. As a result, what happens at a conference cannot be interpreted solely by 'being there'. It must be situated in a larger context of media framings and interpretations that give it significance and meaning and determine its effects. And yet being at a conference does expose the observer to the emotional layer of a COP, from environmental activists' expressions of anger and hope (González-Hidalgo & Zografos, 2020; Ransan-Cooper et al., 2018), to emotional communication by diplomats (Hall, 2015) and NGOs (Salgado, 2018). Experiencing these emotions and reflecting on one's own feelings is an integral part of the ethnographic journey. It complements document- or interview-centred approaches with an additional dimension of sensorial, emotional and bodily experience.

3.3 Spatiotemporal Boundaries and Overflows

Not included in this recurring dramaturgy is the fact that individual conferences usually only punctuate more extended governance processes, acting as the point of culmination of year-long preparations. Preliminary negotiations and intersessional meetings shape the agenda of each conference, determine draft outcomes and decide which topics will be included and which will not. After the conference, its outcomes in the form of decisions, declarations, protocols or media frames are brought into circulation in different national and local contexts, commented on by activists and academics, interpreted by legal scholars, and ratified and implemented by national governments. Hence, while every global environmental conference is a singular event with clear temporal and interactional boundaries, it is also shaped by continual temporal and spatial overflows (Schüssler et al., 2014). Global environmental conferences have a predetermined beginning and end. But negotiations at one COP are also part of a longer history, and are connected to processes outside the conference halls. Temporal boundedness is regularly contested and reaffirmed, for example when the scheduled end of a conference is used to evoke a sense of urgency and put pressure on negotiators.

Collaborative Ethnography of Global Environmental Governance   23

'It is always 5 minutes to midnight!'

At COP21 in Paris for example, the co-chairs of the main negotiation bodies, UN officials and French COP president Laurent Fabius,

> tirelessly repeated that the conference would not be prolonged beyond Saturday night. This created a sense of urgency that was amplified because the crucial importance of COP21 had been almost ritually highlighted by the UNFCCC secretariat and French officials in the run-up to the conference. For example, UNFCCC Executive Secretary Christina Figueres travelled the globe to remind that Paris was the 'last chance' to strike a deal, while media outlets stylised the Paris talks as 'twelve days that will decide Earth's future'. Backed by results from modelling exercises that indicated a rapidly closing window for climate action, alarmism also dominated opening statements. (Aykut, 2017, p. 21)

Hence, temporal boundedness is constructed through scheduling, framing and media coverage. UN conferences are also notoriously prone to last-minute drama, prolongations, and the adjournment of controversial issues to follow-up meetings (see also: Galbraith, 2015; Reychler, 2015). Interactional boundedness is the tendency of UN conferences to segregate social spaces into different meeting zones and to stabilise communication networks among similar kinds of people. Discussion within these circles is often limited to non-controversial topics. The designed spatiality of a conference usually reflect these separations. Scholars of organisation studies (Schüssler et al., 2014) have pointed to the importance of temporal and interactional boundedness in reproducing social order; but they have also highlighted the capacity of conferences to sometimes act as 'field-configuring events' that transform a governance domain. The latter is enabled by interactional openness, where participants are able to mingle and freely interact with a diversity of other interlocutors in informal settings. Interactional openness is crucial to breaking out of the routine functioning of UN conferences and facilitating path departure. It is seriously limited, however, by strict rules of accreditation and social stratification of access, and by budgetary and travel restrictions (Siméant et al., 2015a, p. 16).

The spatial boundaries of a COP or UN summit are marked by the confines of the conference building, complete with security checks and badge controls. But here too we find overflows. Important conferences are usually accompanied by a groundswell of civil society activities within and outside the conference building, as well as happenings in the streets of the host city. Already the 1972 UN Conference on the Human Environment in Stockholm was accompanied by an 'Environmental Forum' and a 'People's Forum' (Hawkes, 1972). At recent UN summits, too, there have been protest event and cultural activities organised by social movement coalitions spread throughout each host city (Chartier & Le Crosnier, 2015). Transnationally coordinated protest actions have also become a common feature of climate COPs, substantiating their character as world events (de Moor et al., 2017). In Glasgow, activists from the COP26 Coalition intended a central space near to the Blue Zone to act as a centralised civil society hub (Aykut, Pavenstädt, et al., 2022). But these plans were thwarted due to delayed construction work just over a month before the start of COP26, resulting in the last-minute rental of various venues. Civil society was thus distributed across Glasgow, with no single place large enough for everyone to gather. Instead, multiple sites acted as key venues, notably Adelaide Place Baptist Church (with a maximum capacity of 220) and the Landing Hub, which was located in the vicinity of the Blue Zone.

The negotiations, too, are not restricted to formal gatherings in the meeting rooms of the conference venue, but regularly continue in informal backroom spaces, the offices of national delegations and the hotels of the host city. Spatial overflow has also increasingly been extending into digital spaces, as UN bodies and summit organisers offer online streams and options for virtual participation. At climate COPs, all plenary meetings as well as increasing numbers of negotiation sessions and side events are streamed on the UNFCCC's digital platform. Social movement assemblies, counter-summit meetings and protest events are also frequently live-streamed, in addition to virtual broadcasts.

3.4 One Event, Many Social Spaces

As nodes in a continuous governance process, global environmental conferences are both diverse ensembles of different meetings and singular events that form a unity (Campbell, Corson, et al., 2014, p. 3). Climate COPs, for instance, are composed of different activities (negotiations, talks and discussions, expert meetings), social spaces (political space, corporate space, activist space, media space), and material locations (dedicated zones and rooms with access restrictions), which partly overlap. Despite differences from one COP to the other, they share common organisational features. As mentioned earlier, these have been described in terms of an internal differentiation into an 'In' of international negotiations, an 'Off' comprising official side events, and a 'Fringe' of separately organised civil society events (Aykut, Pavenstädt, et al., 2022; Dahan et al., 2009).

Three Circles of Global Climate Governance

Every UN conference has a so-called 'Blue Zone', access to which is restricted by a system of badges. Only registered attendees, state delegates or observers affiliated with an accredited organisation are allowed to enter. Registration usually takes place well in advance and places are limited. Badges denote which organisation or country a participant is affiliated with. Within the Blue Zone, there are specific rooms or areas that are only accessible for country representatives, providing them with a protected space for negotiations. Other areas are specifically reserved for UN personnel or media representatives, and some meetings or briefings reserved for specific categories of observers. This system of differentiated access restrictions creates an insider/outsider dynamic regarding flows of people and communications, which greatly contributes to the complexity of the event.

As they do not have such differentiated restrictions of access, events within the large plenary rooms are usually the most 'public' events within the Blue Zone. Seating outside the delegate rows is usually attributed on a first-come-first-serve basis. Adjacent to these ceremonial plenary halls are more sober and informal negotiation spaces, which host the public and non-public sessions of different UN bodies. At larger conferences such as climate COPs, a large part of the Blue Zone is usually made up of country pavilions and observer stands. In this space, which is similar to a trade fair, countries, corporations and international organisations showcase their ideas, publications and best practices, advertise technologies and solutions, and host side events, debates and talks. The pavilions also provide occasions for networking and spontaneous interactions with passers-by. During a conference day, most of these stalls change their social character from information-booth-style PR platforms and miniature conference sessions complete with whiteboards and panel discussions to informal networking hubs facilitated by complementary refreshments and drinks. Other than the pavilions and negotiation space, the Blue Zone generally includes working and meeting spaces for government and non-government representatives. The corridors connecting different parts of the Blue Zone form a distinctive social space in their own right, which may be used by COP Presidencies for messaging, by social movements for protest activities and by reporters as backgrounds for video broadcasting. More informal areas such as cafeterias, cafés, restaurants and sometimes smoking areas provide further networking opportunities. A testimony to increasing media attention, recent climate COPs have also included a 'media hub', used for interviews, live broadcasts and moderated panel discussions. While much of the Blue Zone allows for free passage through its composite spaces, it also acts as a bubble that separates an insider space from the outside. 'The walls here are very thick', said one activist at COP26 in Glasgow, describing a situation where the inside becomes an almost mythical place of projection and curiosity ('What are they doing in there?').

In-between the Blue Zone and the larger city space, COPs and other UN summits sometimes feature a semi-official Green Zone, an 'Off' to the

'In', designated as a space for civil society to hold side events. Generally open to the public, preregistration is nonetheless often required to enter this space. In the climate arena, the Green Zone has in recent years become an increasingly corporate space, to the detriment of environmental NGOs and social movements. The Glasgow Green Zone at COP26, for instance, hosted in the city's Science Centre, boasted a car park filled with hydrogen-powered cars, including a race car, as well as a whole floor dedicated exclusively to corporations presenting their green efforts in modular presentation stalls, like a miniature version of the Blue Zone's pavilion area (Aykut, Pavenstädt, et al., 2022).

As only a small number of participants and activists can gain access to the UN spaces of climate conferences, the streets of the host city are often transformed into 'Fringe' stages for the COP. At more important conferences, this Fringe space may include a more institutionalised civil society space – such as the Cúpula dos Povos at Rio+20 in 2012 (Chartier & Le Crosnier, 2015), or the People's Summit for Climate Justice in Glasgow. These are organised as counter-summits (Meek, 2015). The Glasgow People's Summit was headquartered in a church where events, talks and panels were held throughout the period of COP26. Daily demonstrations in the streets by social movements like Extinction Rebellion and Fridays for Future aimed to create pressure and a sense of urgency, culminating in the Global Day of Action/People's March for Climate Justice on the weekend in the middle of COP26, with over 100,000 attendees. The Fringe spaces throughout the city set up by the Coalition also were heterogeneous both physically and socially, from the more conference-like panel discussions and presentations of the counter-summit to more intimate meetings in the Coalition headquarters, where strategy and networking events took place. However, there also were constant overlaps and spillovers between the In, Off and Fringe, with some actors present in two or more spaces, and some activities repeated with slight variations throughout the larger space of the conference.

The 'Off' and the 'Fringe' beyond Climate COPs

While this structure was first described for climate COPs, biodiversity COPs are structurally similar. With total attendance generally lower, the Fringe in particular is decidedly smaller (Campbell, Corson, et al., 2014). World Conservation Congresses, by contrast, are organised differently, with a temporal rather than spatial separation between different social spheres. Held by the International Union for Conservation of Nature (IUCN), they also include states, government bodies, national and international NGOs and 'affiliates' such as scientists forming 'the world's largest and oldest network of environmentalists, in which the private sector, nongovernmental organisations, governments, and civil society work together to define the conservation agenda' (Doolittle, 2010, p. 288). The Congress begins with a 'Forum' in the first week: this event, which loosely corresponds to the Off of a COP, is a time for different actors to meet and exchange. It has been characterised as 'part conference, part convention trade show' (Gray, 2010, p. 332; see also Brosius & Campbell, 2010). This is followed by a 'Members' Assembly' in the second week, where accredited members vote on motions and negotiate in an atmosphere that has been described as more contentious than the Forum (MacDonald, 2010). Here, the workshop is a characteristic meeting format (Hagerman et al., 2010). Overall, as a non-UN event, there appears to be a lower bar for entry, especially to the Forum, which is intended as a space for public debate, while the Members' Assembly, as a decision-making body, akin to the In of a UN COP, counts state and government representatives among its attendees. The aspect of theatricality and spectacle is also very prominent throughout the different spaces of each WCC (MacDonald, 2010). And even the surrounding civil society activities have been on the rise as of late. A counter-summit was organised two days in advance of the official World Conservation Congress in Marseille in 2021, in accordance with the temporal logic of the main event.

4 Methodological Building Blocks

4.1 Focused Ethnography

Ethnography is traditionally characterised by extensive stays in the field, a practice that has shaped the common picture of ethnographic methods. But global environmental conferences typically last no longer than two weeks (plus the almost customary one or two extra days of negotiations extended into the weekend). This poses a problem for event ethnographers, as their field of study is mobile and short-lived rather than geographically bounded, becoming observable only within short temporal intervals in different locations, for the time of a conference, congress or summit. This problem can be addressed by applying methods from *focused ethnography*. This term was brought into common usage in its current meaning by Knoblauch (2005), who presents it as an extension and adaptation of traditional ethnographic methods. Rather than being in opposition to more traditional long-term fieldwork, he frames it as a research strategy that is 'complementary to conventional ethnography, particularly in fields which are characteristic of socially and functionally differentiated contemporary society' (p. 1). Focused ethnography is first and foremost a pragmatic approach: in research settings where extensive stays in the field are not possible, intensive short-term fieldwork is done instead. This does not mean that focused ethnography is a 'quick and dirty' or 'mini' version of ethnography (Wall, 2015, p. 4). Quite to the contrary, large chunks of research are done outside the field, in preparation for the fieldwork and in subsequent analyses of supplementary data sources, for instance to understand how coalitions are forged and shared meanings are established even before formal negotiations start (Müller & Cloiseau, 2015). Table 1 offers a comparative presentation of key differences between 'conventional' and focused ethnography.

Focused ethnographies have been heavily employed in practice-oriented disciplines such as engineering and healthcare, mostly to analyse work routines and propose improvements (Higginbottom et al., 2013; Millen, 2000; Wall, 2015). These applications of focused ethnography have been criticised for losing core qualities of ethnographic work, such as the deep experience and immersion that only long-term fieldwork can offer, and the ability to capture complex processes of meaning-making by members of a community (Breidenstein & Hirschauer, 2002). As a response to such critiques, Wall stresses the pragmatic aspects of focused ethnography: 'Rather than being a threat to the ethnographic endeavour, focused ethnography preserves the essential nature of ethnography and allows researchers to explore cultural contexts that cannot be studied using conventional ethnographic methods.' (Wall, 2015, p. 5). The key strength of focused ethnography, she adds, is that it enables the ethnographer to study more geographically dispersed and temporally ephemeral communities. When studying a group one is not a member of, important amounts of background knowledge are required. For researchers in a specific domain of global environmental governance, this involves specialist knowledge of the domain's history, its main institutions, actor constellations

Table 1 Comparison between conventional and focused ethnography (Source: Knoblauch, 2005, p.7).

Conventional ethnography	Focused ethnography
Long-term field visits	Short-term field visits
Experientially intensive	Data/analysis intensity
Time extensive	Time intensity
Writing	Recording
Solitary data collection and analysis	Data session groups
Open	Focused
Social fields	Communicative activities
Participant role	Field-observer role
Insider knowledge	Background knowledge
Subjective understanding	Conservation
Notes	Notes and transcripts
Coding	Coding and sequential analysis

and norms, as well as formal and informal rules. It is impossible to pick up such elements in a short field stay. Studying environmental conferences thus requires going into the field well prepared, and possibly returning to it in an iterative succession of short, focused field stays. Focused ethnography is well suited for such an approach: '[F]ocused ethnographies are studies of highly differentiated divisions of labour and a highly fragmented culture. The more diverse and short-term the fields and activities to be observed become, the more flexible, short-term and focused should be the instruments of our research' (Knoblauch, 2005, p. 11).

Some properties of global environmental conferences can even be used to the ethnographer's advantage. As outlined in the previous section, over time UN governance spaces produce a small-world effect, where repeated interactions create in-groups of 'COP-junkies' (Dumoulin Kervran, 2021, p. 91). Such worlds can be successfully observed and analysed only by joining them through a series of repeated short-term observations. As Campbell and colleagues (2014, p. 16) note, their 'familiarity with the policy negotiations at stake (their histories, the various interests groups involved, etc.) rivals that of more traditional ethnographers who understand land tenure or household division of labor in particular places'. COPs' repetitive temporal structure enables ethnographers to go beyond a single well-prepared period of data collection in a given field, engaging instead in a longer-term, iterative research process, and thus in longer-term and more in-depth research projects. This fits with the ethnographic insight that fieldworkers who engage in longer-term field stays not only get to know the field well, but to an extent become part of the field that they study (Hyndman, 2001). The preparation time needed to understand the overall technical complexities of a given field in global environmental governance decreases as it becomes increasingly familiar, and more preparation time can be spent on specific research questions. In the process, conference ethnography can be iteratively refined. This partly makes up for certain shortcomings of focused ethnography, and has been stressed by many researchers (Corson et al., 2019; Gray et al., 2019). It is also an example of a more general practice in ethnographic research in which researchers regularly return to a given field, in order to investigate changes in it over time, and to observe change in the researchers' own perspective on it (Schnegg, 2023, 2021). Repeated research on recurring events like the climate and biodiversity COPs can not only help avoid some of the pitfalls of short-term ethnographic observation, but also respond to some of the questions and problems that arise when ethnographers return to a field in which they previously studied (Ellis, 1995). The aspect of gaining perspective through what O'Reilly has called ethnographic returning (O'Reilly, 2012) can be strengthened by returning to the field in a team, and productively extended to a studying up context. It may be possible to better answer questions on the attribution of observed changes – has the field changed? have I as a researcher changed? – when returning to the field with either the same or different team members.

4.2 Team Ethnography

Collaboration is a key aspect of scientific life, and is often understood to have an advantage over working alone, especially when dealing with complex and fragmentary events (Brosius & Campbell, 2010, p. 248). And yet the stereotypical ethnographer works alone. Especially in the anthropological tradition, where extensive fieldwork is the cornerstone of disciplinary training, a paradigmatic view came into currency that was described, perhaps overpointedly, by the anthropologist Robert Hackenberg as the '1:1:1 ratio – one man [sic], one village, one year' (quoted in Erickson & Stull, 1998, p. 2). However, the 'lone ranger' ethnographer is largely a myth. In their guide to what they term *team ethnography*, the anthropologists Ken Erickson and Donald Stull show how from the very beginning, 'It was the expedition, not the independent investigator, that the first ethnographers chose as their research model when they got up out of their armchairs and went to the field' (Erickson & Stull, 1998, p. 3). Moreover, collaboration frequently extends to informants and research subjects. In recent years, the term 'collaborative ethnography' has thus also been used to encapsulate the ethnographer's collaboration with the people and communities they study (Lassiter, 2005). This approach emerged partly in response to what is known as the 'crisis of representation' in anthropology, which was driven by the realisation that the voices and viewpoints of the people who were the focus of anthropological studies had been excluded from the practice of ethnographic research (Clifford & Marcus, 1986). Rather than telling stories *about* communities and cultures, collaborative ethnography attempts to work *with* those who are studied, in an approach that shares some features of action research or transdisciplinary research. The journal *Collaborative Anthropologies* embodies this quest for new and more ethical research methods (Rappaport, 2008). Clerke and Hopwood (2014, pp. 14–15) choose instead to refer to ethnographic endeavours that involve large groups of researchers as 'doing ethnography together', doing ethnography 'in teams' and 'co-ethnography'. This multiplicity of terminologies stems partly from the multiple types of arrangements that working in a team can involve, ranging also form one-time observations to longitudinal studies. These can include collaboration in ethnographic data gathering and interpretation over an extended period within a sedentary community (Buford May & Pattillo-McCoy, 2000; Gerstl-Pepin & Gunzenhauser, 2002; Low et al., 2005), common observations of a single complex and crowded event site (Mazie & Woods, 2003; Paulsen, 2009), longer-term collaboration involving division of labour to observe a sequence of events (Gray et al., 2019), and parallel work by observation teams at parallel field sites (Jarzabkowski et al., 2015). It is these collective or team-based approaches that inspired early global event ethnographers, as they promise to remediate some of the most significant problems with short-term observation. As Siméant and colleagues note, 'Given the limited time available to researchers and their lack of familiarity with the local context, [observing such events] presupposes engaging in collective observation […] whenever possible by multinational teams to prevent ethnocentric interpretation' (Siméant et al., 2015a, p. 13). If practised effectively, collaboration among larger teams of researchers can help avoiding 'the pitfalls of "airport" or "parachute" ethnography' such as 'minimal first-hand observation or over-reliance on institutional records, media accounts, or memories' (Paulsen, 2009, p. 510).

But teamwork is always a balancing act. According to Erickson and Stull, there are no easy recipes for team ethnography, as 'research goals and contexts will largely dictate whose company the ethnographer keeps' (Erickson & Stull, 1998, p. 5). They go on to argue that 'if we really think about it, if we are really honest with ourselves about what it is we do, the question for the ethnographer becomes not whether to team or not to team; ethnography is by its very nature a team enterprise. The question becomes, What do we want our ethnographic team to look like? Whose understandings shall we include?' (Erickson & Stull, 1998, p. 59). However, many pieces on team ethnography contain sections on disadvantages and difficulties, or end with 'warnings' of how and when not to practice the method. Some authors have cautioned, for example, that rigid and inegalitarian division of labour in the research process can deprive ethnography of its main advantages and strengths, namely immersion, a holistic approach to the research process, and an in-depth understanding of local meaning-making (Mauthner & Doucet, 2008). Citing Platt's (1976) dictum that 'knowledge once divided can be hard to put together again', Mauthner and Doucet (2008, p. 976) plead against the assignment of fieldwork and data analysis, or of research management and implementation, to separate team members, stressing that the embodied and immersive aspect of fieldwork is an essential part of the ethnographic method. Theirs is not a dismissal of team-based ethnography as such, but a warning against hierarchical and unequal forms of division of labour in neoliberal, production-oriented research (re)organisation that have been criticised in the name of 'slow scholarship' (Mountz et al., 2015).

Various groups of researchers have used team-based approaches in varying configurations to conduct ethnographic observations at environmental conferences. One key aspect of team research on recurring events like climate and biodiversity COPs is the benefits of iterative refinement with either the same or different team members. We discuss some practical aspects, benefits and challenges of this research strategy in greater detail in Section 5.

4.3 Digital Ethnography

Before the COVID-19 pandemic, the United Nations General Assembly had scheduled the final negotiations for the ratification of an intergovernmental treaty aimed at establishing binding rules for marine conservation to take place in the summer of 2020. Negotiations and preparations for this treaty had been over a decade in the making (Vadrot et al., 2021). The pandemic forced the postponement of the summit, but negotiation spaces were opened in the digital sphere to avoid losing the momentum that had been built within the process. This forced the observing researchers to shift their focus to these digital spaces. While the shockwaves sent by the COVID-19 pandemic through the world of international conferences and meetings appear to be incrementally subsiding, the foregrounding of digital spaces that they catalysed continues, and researchers working ethnographically at international conferences must adapt.

This is not entirely new. Since its inception in nineteenth-century anthropology, ethnography has continuously adapted its main approaches and central concepts, including those of the field and fieldwork, to social changes and new developments. This remains the case today, in a time when one of the most dramatic transformations in life-worlds has been the emergence and continuous spread of the digital sphere. This has not been lost on ethnographers (Markham, 2013). The question of how to do ethnography in an increasingly digital age has been answered in various ways. Early studies with an 'ethnographic approach to the internet' and the use of information and communication media (ICM) made use of media ethnography, studying the offline communities that have emerged around social platforms and online sites (Miller & Slater, 2001). In a similar vein, virtual ethnography (Hine, 2000) has emerged as a means to study new communities and forms of interaction in online communities. But the approach closest to the concerns of scholars interested in studying global environmental conferences is that of *digital ethnography* (Abidin & de Seta, 2020; Hsu, 2014; Pink et al., 2016). Such research has been described as moving from studying the Internet as a medium in its own right towards using the Internet to study culture and society (Rogers, 2009, p. 29). Digital ethnography takes as its starting point the observation that the distinction between online and offline is not straightforwardly marked, digital communication, for example through social media, is part of people's everyday life and activities. Postill and Pink (2012) call this the 'messy web'. Starting from a critical media studies perspective, digital ethnography takes a non-media-centric and non-digital-centric approach to the digital. For example, it does not assume Twitter is any more perfect a representation of contemporary societal structures and trends than newspapers, television, or any other popular medium is, or has been. But using Twitter to study society can highlight novel aspects of contemporary life. It is the aggregate and productively contradictory picture that emerges from a combination of observing developments in all of these different media which is ultimately of the greatest value.

A key challenge in the online world is the lack of face-to-face, physically co-present interaction. As interpersonal skills and interactions are not the centre of attention, digital ethnography typically focuses on textual or video analysis. The forms of data gathered, including screenshots and recordings, also differ from those that feature in offline ethnography. The role of recordings highlights an affinity to focused ethnography. While working with recorded material poses problems of temporality and context, a benefit is that digital methods may span multiple scales, making it possible to zoom in and out (Hsu, 2014).

With 'lurking' as a common practice in digital ethnography (Ferguson, 2017), practical, theoretical and ethical issues arise around the question of 'What is participation?' What is it that is observed, what does participant observation mean, what is the field, what kind of data are we able to work with, and what kind of data is it acceptable to work with? Markham (2013) pointedly notes that all of these questions are already present in traditional ethnography – and that far from having been conclusively answered, they remain part of ongoing debates and reflection, keeping pace with new contexts of ethnographic study, just as in ethnography's early days. The question of what the field is cannot be conclusively answered, and is not subject to abstract, universal delimitation, but instead depends on the research topic (Garcia et al., 2009). Overall, and despite the obvious existence of differences, many of the same rules apply in digital as in non-digital ethnography, including the value of mixed methods approaches and a role for autoethnographies.

Like other social scientists keeping pace with the expansion of digital spaces in their respective fields, ethnographically oriented researchers of global environmental governance have added digital methods to their repertoire of approaches in recent years. Although digital spaces have been steadily attracting more activity within environmental summitry – official negotiations, media, observers and protesters – the COVID-19 pandemic propelled virtual meeting spaces and online communication to the centre of attention (Chasek, 2021).

Already before the pandemic, websites, social media and virtual discussion platforms were increasingly becoming sites of contestation. Marion Suiseeya and Zanotti (2019), for example, note the charge that the COPs were inaccessible for representatives from the Global South and Indigenous activists and representatives, with both physical attendance and the structures of participation described as exclusionary. This had spurred activists' participation in online events, such as at COP21. Although activities in virtual fora might be thought of as a sideshow to the main, physical COP events, Marion Suiseeya and Zanotti show that online activism played a crucial part in the push to include some of the proposals and demands of Indigenous peoples' groups.

Since COVID-19, digital spaces have become even more important, and are coming increasingly into view as sites not just of communication and contestation but of negotiation. Vadrot and Ruiz Rodríguez (2022) point out that online platforms are increasingly being used for diplomatic negotiations, and that what is happening there is not yet well understood. Their digital ethnography of international online negotiations on marine biodiversity highlights the emergence of what they call 'digital multilateralism'. While many of the traditional diplomatic practices are also present online, the face-to-face interactions and informal 'coffee talks' characteristic of traditional diplomatic negotiations – which, some argue, are of fundamental importance – are not. But some new practices have also emerged that are exclusive to (partially) digital negotiations processes, and that in the future will perhaps become more widespread. Vadrot and Ruiz Rodríguez (2022, p. 11) point to the intertwining of online and offline negotiations as a case in point. While some have welcomed the expanding role of online fora as an opportunity for increased inclusivity as more actors are able to participate in events, decisions are still made at in-person negotiations by quite a small number of actors. In the process, both online and offline negotiations become increasingly difficult to follow for observing ethnographers and participants alike, without the information and background that are discussed in complementary in-person and virtual spaces.

4.4 Dramaturgical Perspectives

Ethnographic accounts of global environmental conferences have traditionally placed a strong focus on performative aspects of these events (Death, 2011; Little, 1995). A dramaturgical perspective is a common thread running through many team ethnographies of global environmental governance, although not always in an explicit form. Dramaturgy was not the focus of the Duke CEE project at its beginnings. But the politics of performance and the staging of the conservation congress itself soon emerged as strong and salient features of the

Collaborative Ethnography of Global Environmental Governance 35

process that they were observing (Brosius & Campbell, 2010, pp. 249–250). The performative dimension of global environmental conferences also figures as a central theme in the French tradition of global event ethnography (Foyer et al., 2017, p. 5). Subsequently, it became a conceptual anchor for understanding global climate politics (Aykut et al., 2021), and has been positioned as a theoretical alternative to the notion of *orchestration* in attempts to understand soft coordination in global climate governance (Aykut, Schenuit, et al., 2022). Various studies on negotiations and official conference activities, or investigating the ways in which social movement actors approach environmental meetings, have focused on symbolic action, role taking, dramaturgy, performance and other theatrical metaphors.

Perhaps in equal parts a methodological building block (Figure 3) and a conceptual framework, a dramaturgical perspective looks at various aspects of performances on the stages of climate conferences. Dramaturgy, with its theatrical metaphors, lends itself especially well to ethnographic observation at these sites. Although talk of performance in relation to people's behaviour in everyday contexts has a mostly negative connotation, for sociologist Erving Goffman (1959, 1963) it is a constitutive – and thus unavoidable – aspect of social interactions. Goffman defines performance as 'all the activity of an individual which occurs during a period marked by his continuous presence before a particular set of observers and which has some influence on the observers' (Goffman, 1959, p. 32). When there is an impression to be made, that is, whenever there is an observer present, self-presentation is necessarily managed. Goffman further speaks of social interactions in terms of a *front* and a *backstage*. The front stage is the place of impression management, where a 'mask' is put on for an audience, which is often aware of the performance and the public character of a situation. Backstage, performers are hidden from the

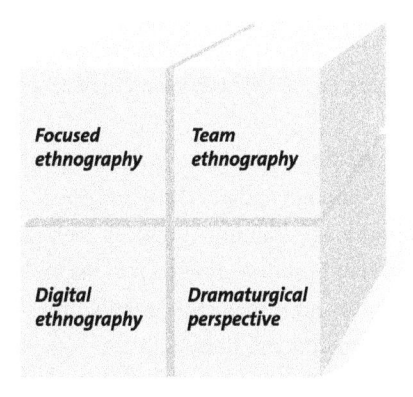

Figure 3 Methodological building blocks of collaborative event ethnography

audience. This is a place of retreat and preparation for future public performances. Front and backstage need not necessarily be physically separated spaces: they are first and foremost social spaces and analytical categories.

A key aspect of the dramaturgical perspective are the various pointers, conventions and guides that shape the way performances take place. In a dramaturgical vocabulary, these are *scripts*, *repertoires* and *roles*. There has been considerable scholarly debate around how they come about, how fixed they are, how they can change, and how people choose between them. Political sociologist Maarten Hajer (1997, 2009) has applied a dramaturgical perspective to public policy discourses in environmental and climate governance. He shows that, particularly in a highly mediatised world, public policy discussions feature role taking and the deployment of repertoires and scripts. He explores issues such as settings and staging within policy and governance processes, roles that claims-makers take in debates, and how repertoires travel. This approach thus aims to analyse the symbolic and discursive aspect of politics (Edelman, 1985).

One approach to social movements and contentious politics that has been widely influential is Charles Tilly's conceptualisation of scripts and repertoires (Tilly, 2006, 2008). It is part of a wider body of work showing that a dramaturgical focus can be fruitfully applied in the context of longer-term observation of the emergence of traditions and ritualisations in national protest cultures (Chaffee, 1993). This has included conferences and large events, and especially the World Social Forum, as an event held by and for social movements (Herrera et al., 2015; Rucht, 2011; Siméant et al., 2015b). The presence of movements at global summits has produced repertoires of counter-summit protest tactics, with a strong emphasis on dramaturgy directed towards the media, making such protests useful mobilising tools for social movements (Juris, 2008). However, presence and protest at summits puts movements at risk of themselves getting caught up in the logic of 'summitry' and 'summit theatre' (Death, 2010, 2011), and even giving rise to a 'counter-summitry' (Meek, 2015). Tensions between varying and potentially contradictory models for action structure a range of roles that protesters may take on at environmental summits.

In ethnographic observation at COPs, dramaturgical analysis provides a useful reference frame, although to date it has rarely been explicitly deployed. The arena of high diplomacy and summitry presents itself as a prime example of staging and self-presentation in action, as shown in the case of the negotiations for the extension of the European Union (Schimmelfennig, 2002). A number of studies have brought a dramaturgical perspective to bear on climate and other conferences, drawing on ethnographic observation and influenced by the 'ethnographic turn' in IR (see Section 2). A central concern here is to analyse practices of soft coordination and orchestration, which occur despite sometimes heated opposition and conflict between different actors, and their interests and views. MacDonald (2010), drawing on observations at World Conservation Congresses, argues that, over the course of many of these events, the concept of conservation itself has been resignified in ways that are beneficial to the business participants in the negotiations, through repeated practices of script development and re-scripting. Speaking of the WCC as a spectacle, he argues that the meeting format employed there has been essential in enabling corporations to subtly shift and reconfigure the topic at hand. 'Events like the WCC constitute the political sites where much of that reconfiguration is rendered legible; where the political future of conservation is negotiated; and where struggles over deciding what binds "us" all together are acted out' (p. 271). Dramaturgical perspectives drawing on Goffmann, Hajer and others remain an underlying focus in many collaborative event ethnographies.

5 Practising Collective Research

Practising collaborative event ethnography is 'easier said than done' (Brosius & Campbell, 2010, p. 247) and sometimes 'more hands do *not* make light work' (Gray et al., 2019, p. 15). Erickson and Stull's *Doing Team Ethnography: Warnings and Advice* (1998) is a valuable survey of some of the recurring issues in collective research, its advantages, and things to look out for before embarking on this complex and challenging endeavour. The authors identify a series of overarching factors that make team-based work in ethnography difficult. These include the cult of individualism in academia and the professional competition built into the academic reward system, the risk of teams splitting into different factions along disciplinary lines or forming subgroups that reflect professional hierarchies, or work being distributed along gendered lines. Considerate and thoughtful organisation is thus at the core of collaborative research. On the other hand, collaborative research also requires a willingness to experiment (Brosius & Campbell, 2010, p. 248), and an approach that is not rigid but 'dynamic and relational' (Corson et al., 2019, p. 59). A broad but fragmented literature on ethnographic methodology provides tips on how such collaboration can be made successful (for an overview, see Clerke & Hopwood, 2014). In addition to that literature, this section draws on our own experience and on the methodological sections of publications on collective ethnographies of global environmental conferences (see especially Brosius & Campbell, 2010, pp. 247–249; Campbell, Corson, et al., 2014, pp. 9–12; Corson et al., 2019, pp. 59–63; Dumoulin Kervran, 2021, pp. 87–91; Foyer, 2015b; Foyer et al., 2017, pp. 11–14; Foyer & Morena, 2015; Gray et al., 2019, pp. 9–16). The key themes that emerge roughly follow the arc of an ethnographic project, from preparing together before the fieldwork begins, to organising collective observations and data gathering, cooperating on data analysis and interpretation, and finally collaborative writing and publication strategies. Some of the themes raised are specific to short-term ethnographies at complex multifocal events like global environmental conferences, and thus differ from issues that can arise in longer-term field stays. Others, relating for instance to preparing fieldwork or writing a joint publication, are common issues in ethnographic team research. The following sections aim to synthesise advice and warnings found in these literatures, selecting what we have found most useful in our own work on climate COPs. Each section ends with a summary list of 'Dos' and 'Don'ts' which are brought together in Table 2. To preserve the ethnographic core of the research practice, the Don't of avoiding rigidity in preparing, working, thinking, being and writing together consistently stands out as a key caveat.

5.1 Preparing Together: Creating Ownership

A key element of focused ethnography is preparation. The fact that observation time is necessarily limited does not mean a shorter research project. Much to the contrary: the lion's share of work shifts to the preparatory phase and to subsequent analyses. Global environmental meetings use highly specialised language and involve complex entanglements of actors and institutional arrangements that act as hurdles to participation. Accordingly, research usually does not start when one arrives in the host city or enters the conference centre, but weeks and months before. It is essential to come to the event prepared, considering that global environmental conferences usually are not singular events, but part of longer cycles with a history and context, with which researchers need to familiarise themselves well before the event starts. Preliminary workshops, seminars and meetings can be used to this end, as can the study of preparatory reports and documents, conducting scoping interviews and contacting informants. Important preparations may also include securing accommodation, getting to know commuting times, public transport timetables and routes in order to minimise the amount of scarce, valuable time in the field that is spent on logistic coordination. Preparatory workshops are also instrumental for attracting interested researchers, putting together a diverse research group with complementary skills and backgrounds, and developing a shared conceptual focus or anchoring concept for use in coordinating work within the team.

Most accounts of collaborative event ethnography stress the importance of the preparatory phase, which includes activities as diverse as setting up a website and a research infrastructure, discussing ethical protocols, inviting external researchers (Brosius & Campbell, 2010, p. 249), holding monthly

Table 2 Dos and Don'ts of practising collective ethnographic research

	Preparing together	Observing together	Thinking together	Experiencing together	Writing together
DO	• deal together with questions of access and accreditation for the whole group • create joint problem ownership for the project • find common interests, develop a team spirit and common research culture • create a common methodological perspective or conceptual vantage point • address issues of unequal access to international arenas	• define a mode of collaboration adapted to the project • organise regular meetings and define daily meeting points • use a common format for observation notes, establish data sharing infrastructure and routines	• create opportunities to discuss and resonate together • find a common conceptual or methodological anchor to facilitate exchange and focus observations • nurture the standing seminar effect through regular meetings and a shared space • openly discuss questions of intellectual ownership	• create a sensitive and caring atmosphere, with moments of conviviality and possibilities to exchange informally • be approachable and supportive as a supervisor or team leader • reflect on power imbalances and unequal distribution of work	• start joint writing projects soon after the event • convey the richness of the field while also connecting it to its context and social science debates • clearly discuss publication strategies and issues of authorship • leave room for individual publication projects • support early career researchers • nurture long-term relations

Table 2 (cont.)

	Preparing together	Observing together	Thinking together	Experiencing together	Writing together
DON'T	• come unprepared expecting to simply see 'what happens' • lose track of individual perspectives and research interests • prematurely define rigid responsibilities and foreclose alternative conceptual perspectives	• establish a fixed, rigid division of labour that takes away spaces for reflection and possibilities for surprise • lose track of the group and its common focus	• preclude observations by prematurely formulating strong hypotheses or defining rigid analytical frameworks	• discard unwanted feelings and emotions • disregard individual needs • uphold strict hierarchies • reproduce relations of domination along class, colonial, racial and gender lines	• lose the richness of the field or erase your positionality when writing • let the event's momentum go to waste when returning home • lose track of your own career in the collective project

project meetings and research seminars, and creating a mailing list and file hosting system to share relevant information (Foyer et al., 2017, p. 13). Ultimately, a core aim of these preparatory activities is to build joint problem ownership (Schuck-Zöller et al., 2018, p. 114) and a shared research culture (Foyer et al., 2017, p. 13), rather than prematurely defining rigid responsibilities and strict work routines. Importantly, participants need to find and nurture their own motivation to participate in the collective research endeavour, rather than being assigned to a task through their hierarchical superior.

For our observations at COP26 in Glasgow, for example, the team included a group of social movement scholars as well as a heterogeneous group of climate governance scholars with different thematic foci. Preparatory workshops were instrumental in bringing the two groups together by defining the contours of a dramaturgical perspective as a joint analytical framework for all sub-projects. We conceptualised Glasgow as composed of a range of stages for public performances, including front- and backstages, and featuring a variety of dramaturgical practices. We also agreed on a schedule for travel and observations, which included distributing accreditations and rooms in shared rented flats in Glasgow, to guarantee individual researchers' access to their particular field, but also adequate coverage of the conference's different physical and social spaces for the project as a whole.

Throughout these preparatory activities, it is important to keep in mind that collaborative event ethnography is a highly unequal research practice. Hurdles to participation are not evenly distributed along lines of gender, class, race or geographic origin. The location of meetings in different countries around the world, their timing in the middle of the academic semester and the fact that intermediary meetings are sometimes held at short notice, represent challenges for all researchers, but especially for early career scholars without adequate funding, scholars in the Global South who may have difficulty obtaining visas, and scholars with responsibilities in care work. Keeping these inequalities in mind and attempting to resolve them as a group is an important part of creating a diverse research group, a good working atmosphere and team spirit. Accreditation to UN conferences is another equity issue. At climate COPs, observers have to be nominated by accredited organisations, and getting such accreditation itself takes time. Accreditations fall into one of nine 'constituencies' officially recognised by the UN system. For academics, this is usually the Research and Independent Non-Governmental Organizations, or RINGO, constituency. It is not unusual for researchers whose organisation is not accredited to mobilise personal and professional networks to obtain a nomination through another accredited organisation. The inequalities created by this system can be somewhat attenuated by showing solidarity with less fortunate colleagues, either by reserving a nomination for someone who would otherwise not have access to one, or by participating in broader pooling systems that allow for broader and more diverse participation in UN conferences.

...

DO: Deal together with questions of access and accreditation for the whole group; create joint problem ownership for the project; find common interests, develop a team spirit and common research culture; create a common methodological perspective or conceptual vantage point; address issues of unequal access to international arenas.

DON'T: Come unprepared expecting to simply see 'what happens'; lose track of individual perspectives and research interests; prematurely define rigid responsibilities and foreclose alternative conceptual perspectives.

5.2 Observing Together: Many Eyes

To maximise observation time during a global environmental conference, it is advisable to arrive at the event site before the event starts, to get acquainted with the various spaces and settle in. Once in the field, the process of defining modalities of collaboration and developing working routines takes centre stage. Shared observation is very advantageous for transnational mega-events like UN climate conferences, but ethnographers practicing participant team observation in other contexts have also highlighted its advantages (Paulsen, 2009). 'Many eyes' see more than two, and people from different backgrounds and with different disciplinary training and research experiences will notice different things and likely complement each other (Mazie & Woods, 2003, p. 30). To realise this potential, team ethnography has to allow for some degree of freedom, accommodating and nurturing individual motivations to participate in collective work while also keeping individual researchers from simply doing their own thing – a problem that has been fittingly likened to 'herding cats' (Erickson & Stull, 1998). Informal hierarchies often emerge in the course of collective work, and have to be dealt with, especially when they challenge formal roles or work routines, or reproduce problematic power structures. Problems of authority are commonplace, as formal leaders may be reluctant to lead, while others may be unwilling to be led.

To avoid some of these problems, clarifying modalities of collaboration can be useful. Gray and colleagues (2019, p. 2) juxtapose two ideal-typical modes: one that is based on a division of labour among individual researchers, and a mode of strong collaboration that emphasises the co-production of research as a team. Both have their advantages and pitfalls. On the one hand, a certain division of roles is practical for analysing complex events like global environmental conferences. With a division of labour, management and leadership aspects come to the fore: 'For a team leader, project management draws effort away from research in the field. Time spent by the team leader in solo fieldwork amounts to time not spent on preliminary comparisons among team-collected data, team meetings, and getting to know the team itself.' (Erickson & Stull, 1998, p. 17). However, while it is difficult to avoid having a certain hierarchy in collective research, collaboration does not have to involve strong central direction. Too strict a division of labour can even be counterproductive given the mode of knowledge production in qualitative research, where data is always embedded in tacit and embodied knowledge about social situations and cultural context, and cannot be easily separated from the researcher who gathered it (Mauthner & Doucet, 2008). Our experiences indicate that cooperating on the basis of individual autonomy, although it is more complex to implement, can be very rewarding in terms of research results.

Whatever the mode of collaboration, it is necessary to establish work routines, including frequent meetings of team members who are working closer together and regular meetings of the whole group. Ideally, these meetings serve to compare notes, share observations and enable the circulation of information about past and upcoming events. Most teams also develop a common observation matrix or form, that reflects common research interests and sometimes a shared conceptual focus, to facilitate data sharing among team members (Campbell, Corson, et al., 2014, p. 11; Foyer et al., 2017, p. 14). Finally, the spatiality of research collaboration matters. In larger, more complex settings, individual researchers may 'get lost' in the event, and the busy schedule may prevent regular meetings. A technique that has proven useful in this context is to work in pairs for observations and interviews. Observation pairs can allow for more complete coverage of very 'thick' events such as protests. In some cases, there can be a safety element to working in pairs and sharing information with the group (Maclin, 2010). If composed with a view to diversity, they can also add cultural sensitivity and reflexivity to the research (Siméant et al., 2015a). Another technique that can be used is to create subgroups that cover different social spaces at a conference (Foyer et al., 2017). This allocation can broadly follow the inside/outside divide, with one group focusing on social movements and activists, the other on the corporate and political spaces within the conference (Aykut, Pavenstädt, et al., 2022). Subgroups can also be formed according to shared thematic interests, such as an ocean team (Campbell et al., 2013) an agriculture team (Demeulenaere & Castro, 2015) or a business team (Benabou et al., 2017). To ensure communication between subgroups, creating shared research data infrastructure is crucial. This includes cloud services for sharing pictures, fieldnotes and documents. Actors at most conferences produce a steady flow of documents, and attempting to follow it can quickly become overwhelming. It is therefore helpful to distribute and rotate the tasks of reading and summarising newsletters, daily bulletins, briefings and the like. To some extent, this work can be supported and facilitated by members of the team who stay at home, follow events online and summarise media coverage and social media.

...

DO: Define a mode of collaboration adapted to the project; organise regular meetings and define daily meeting points; use a common format for observation notes, establish data sharing infrastructure and routines.

DON'T: Establish a fixed, rigid division of labour that takes away spaces for reflection and possibilities for surprise; lose track of the group and its common focus.

5.3 Thinking Together: Many Minds

Another key advantage of collaborative work that many practitioners highlight is that 'many minds' have more ideas than one (Brosius & Campbell, 2010, p. 248; see also Crow et al., 1992). The founders of collaborative event ethnography, for instance, point out that they started out with an individualistic and pragmatic approach of combining 'many eyes' to more fully capture a complex event, but quickly realised during the research process that regular discussions among team members also brought out insights that would not have emerged in individual research (Campbell, Corson, et al., 2014, p. 10). '22 Heads (and Bodies, and Digital Records) are Better than One', they note (Brosius & Campbell, 2010, p. 248). Each individual researcher brings their own experience and prior knowledge, especially when they come from multiple disciplinary and/or cultural backgrounds (Foyer et al., 2017). However, achieving intersubjective understanding can also be difficult and time-consuming (Gray et al., 2019, p. 12). Sharing data and discussing observations takes time and creates opportunity costs, especially in the context of limited time capacities during fieldwork. In other words, there are trade-offs when moving from loose collaboration among individual researchers to strong collaboration aimed at embarking on a shared intellectual journey.

Among the key factors that make team ethnography difficult, Erickson and Stull (1998, Chapter 3) cite the cult of individualism prevalent in academia – perhaps especially in ethnography (see also Corson et al., 2019, p. 59) – and professional competition, which is connected to the academic reward system. Instead of harnessing the advantages of interdisciplinary complementarity, ill-conceived collaboration can therefore lead to a scattering of team members, or the loss of team focus through the pursuit of individual projects. One way to navigate these trade-offs is to find a common conceptual or methodological anchor that keeps the team focused while allowing for sufficient individual autonomy. Examples cited in the literature include the concepts of scalar differences (Campbell, Corson, et al., 2014), assemblage (Corson et al., 2019) and climatisation (Aykut et al., 2017). In our observation at COP26 in Glasgow, we used a dramaturgical understanding as a shared analytical and methodological frame, which allowed us to conceptualise Glasgow as a series of stages for multiple actors to perform upon, from government representatives, businesses and think tanks at the official conference negotiations, side events and high-level events, to dramatic activist performances in the streets of Glasgow. This dramaturgical approach served both as a broad lens that accommodated different theoretical perspectives (Goffman, Hajer) and thematic interests, as well as a methodological common ground for conducting observations and orienting interview questions.

On a more pragmatic note, the creation of a common discursive space is greatly facilitated by the existence of shared physical spaces that allow for regular and sometimes unplanned discussions, such as a shared flat or daily commutes and train rides. Additionally, regular meetings should be organised on-site in the form of brief follow-up discussions after an event, in shared coffee breaks or meals, and after the event, at the restaurant or over drinks. Continuous exchange and discussions create a 'permanent' or 'standing seminar effect' (Dumoulin Kervran, 2021, p. 89; Foyer, 2015a, p. 19) that facilitates the emergence of novel ideas and creative associations. Developing new ideas during a collective observation also means that they can be put into practice or tested immediately or the next day. One caveat should be mentioned, however. Collective work and discussions also mean that it can sometimes be challenging to clearly attribute an idea or a concept to a particular researcher. Collective work to some degree implies collective ownership of ideas (Gray et al., 2019, p. 14). It is therefore important to create an atmosphere in which questions of intellectual ownership and ways of citing concepts can be addressed and clarified before publication. Partly for that reason, but also to facilitate work, it can sometimes be preferable to disperse parts of the research collective into smaller, more cohesive units with some degree of autonomy.

...

DO: Create opportunities to discuss and resonate together; find a common conceptual or methodological anchor to facilitate exchange and focus observations; nurture the standing seminar effect through regular meetings and a shared space; openly discuss questions of intellectual ownership.

DON'T: Preclude observations by prematurely formulating strong hypotheses or defining rigid analytical frameworks.

5.4 Experiencing Together: Reflexivity and Conviviality

While this is true of other qualitative research approaches, ethnographic observation in particular involves the 'whole' researcher – participating and observing, seeing and listening, but also sensing and being there in the situation. Recent rounds of reflexive discussion on the approach have centred on this aspect of ethnographic fieldwork (Honer & Hitzler, 2015; Spittler, 2001), which sometimes remains in the background in studies of collaborative event ethnography. This may be in part due to the difficulties of treading the path between representing the different field experiences of many researchers while still emphasising the collaborative character of the research. While aiming to avoid the pitfalls of 'confessional tales' (Van Maanen, 2011), here we will discuss some aspects of 'being there' that are particularly relevant to the field sites of collaborative event ethnography.

The importance of bodily, sensory experience in the field relates, first of all, to observations. For example, moments of unanticipated crisis may erupt in a negotiation and manifest in representatives' body language and expression (Hughes & Vadrot, 2019). Such moments provide ethnographic researchers with important insights, which would be impossible to harness without 'being there' (O'Neill & Haas, 2019, p. 8). But bodily experience is not only about observing and sensing the signs of others' states and (re)actions; it includes the ethnographer's own lived experience. Global environmental conferences can estrange, astonish or overwhelm when experienced for the first time. Such emotions can be a powerful resource for analysis, as they open new angles for reflection. However, they can also be an emotional burden for the individual researcher, for example as they realise how hard it is to get access to people in a busy conference setting, or when fear of missing out kicks in, as it becomes clear how little of such an event it is possible to attend. In both cases, working in a team can be beneficial. Seeing the field through the eyes of a newcomer can sensitise the seasoned event ethnographer to the strangeness and peculiarities of a place they have come to think they know all too well. And when anxiety, anger or frustration become overwhelming, experienced team members can help navigate these emotions and put them in perspective. Collaborative event ethnography can also be difficult on a more personal, or intimate, level. We all have our political and ethical convictions, and more often than not, researchers working on global environmental governance will also have sympathy for the causes of stopping environmental destruction and protecting basic human rights. They may find it hard to accept failure or stalemate in negotiations, or feel appalled by displays of cynicism and greenwashing. While observing activists or hearing dramatic speeches, some will be confronted with their own climate anxiety (Clayton, 2020). Here, collaborative work also means being attentive to oneself, and respecting boundaries while being supportive of one another. Functioning teams are not only work units but also networks of support, which offer team members the opportunity to share concerns, doubts and uncertainties.

This is especially true when an essential ingredient is added to the collective experience: conviviality (Foyer et al., 2017). Accounts of successful collaborative event ethnography often highlight the importance of sharing informal moments to create interpersonal ties and even friendships, be it through the sharing of significant research moments, intellectual excitement and camaraderie (Gray et al., 2019, pp. 10–11), or late-night dinner and tapas in a lively Barcelona district (Brosius & Campbell, 2010, p. 248). That being said, we should avoid painting too rosy a picture of academic collaborations. Hierarchies in academia frequently reflect broader societal inequalities, power imbalances and forms of discrimination, and it is important to institute safeguards against discriminatory and abusive behaviour. Thus, collaborative event ethnographers have noted gender imbalances in the team, as 'women have tended to share more, and richer, data; women have been more willing to perform team tasks vs. individual tasks (for example, collect data for collective projects vs. individual projects); and women have performed more tedious, logistical tasks associated with conducting fieldwork (completing ethical applications, arranging access, booking tickets, finding accommodations, coordinating fieldwork activities)' (Gray et al., 2019, p. 13). It is thus important to reflect, discuss and manage gendered divisions of labour and imbalances in collective research practices. This also applies to research collaborations in the field. When working with activists or Indigenous communities, for instance, it is crucial not to reproduce extractive systems of knowledge production. This entails reflecting on gendered and racialised power dynamics and colonial legacies, and attempting to be relevant to those we collaborate with in our research (Wilkens & Datchoua-Tirvaudey, 2022).

...

DO: Create a sensitive and caring atmosphere, with moments of conviviality and possibilities to exchange informally; be approachable and supportive as a supervisor or team leader; reflect on power imbalances and unequal distribution of work.

DON'T: Discard unwanted feelings and emotions; disregard individual needs; uphold strict hierarchies; reproduce relations of domination along class, colonial, racial and gender lines.

5.5 Writing Together ... and Alone!

Ethnography's core strength lies in the in-depth study and thick description of meaning-making at a specific site. This allows attention to be drawn to seemingly mundane things such as waste-management (Herrera et al., 2015), the material setting of a conference (Dumoulin Kervran, 2015) or the design of its civil society spaces (Aykut, Schenuit, et al., 2022). When writing up your ethnographic account, it is important to render this richness and thickness of description. Moreover, the ethnographers should weave their positionality in the meaning-making, by alternating intersubjective and subjective perspectives, and infusing their positionality throughout the written account, not only as an add-on at the end. The description should not be sanitised, and fieldnotes can be used throughout to describe not only what was said and what things looked like, but also what it felt like to being there. Finally, it is also important to situate what has been observed in the field into its broader social and political context, and to embed the insights within the broader social science literature. To make the meaning-making at a site relevant to broader social and political questions, it can be useful to go back-and-forth between ethnographic fieldnotes and broader academic debates.

While these are general traits of all kinds of ethnographic research, the writing phase in collective ethnographies poses a series of supplementary issues and questions. First, while individual ethnographers can take their time to digest after an event and slowly work through their fieldwork notes, this is more difficult in a group, as the momentum of being and working together quickly vanishes after the event. People might return home to different places and reconnect with their personal and professional environments, where they are confronted with other priorities. It therefore appears crucial to reconvene as soon as possible after the event for a coordination of writing and publication strategies. Besides punctual seminars and webinars, longer retreats can be particularly helpful in the collective writing process (Campbell, Corson, et al., 2014). As Dumoulin Kervran (2021, p. 90) writes, 'intensive writing seminars in the countryside proved to be very productive for the re-elaboration of our common framework and fueled the enthusiastic spirit of the team'. Working towards a common publication in the form of an edited volume, a special issue or a report helps drawing things together and making sense of the event as a whole, beyond individual accounts. However, collaboration should not preclude individual publication projects. Especially early career researchers should not 'burn' all their ideas in a common publication, but should be encouraged instead to use parts of their observations for individual purposes. Collective publication projects may also be a difficult sell to funding agencies and publishers. Writing in the 1990s, when collaborative research projects were even less common, Erickson and Stull (1998, p. 49) warn, 'agencies are not interested in polyvocality […]. Book publishers are not much different. Most are not interested in edited volumes, and even fewer want ethnographies that look like them'.

While these latter circumstances might have changed somewhat, negotiating and choosing among different publication strategies is still a crucial aspect of the post-fieldwork part of collaborative event ethnography. Gray and colleagues (2019, p. 14) also point to problems of authorship. For example, being part of a multi-authored publication is not valued in the same way in different disciplines. Departmental expectations about sole authorship might vary and place additional burden on some researchers (Campbell, Corson, et al., 2014, p. 17). Authors might also be disappointed with their position in the author list of a publication. To successfully navigate these issues, it appears crucial to clearly discuss publication outlets and authorship strategies. Innovative formats such as collective authorship might seem appealing, but are not always feasible, especially for non-tenured members of a team (Gray et al., 2019, p. 14). It is equally important to acknowledge those who were involved in broader conceptual conversations either through co-authorship or within the text, and to leave room for individual publications, especially for early career researchers. More broadly, collective ethnographies also make it necessary to define rules for the use of data and acknowledge when observations by a team member are used.

Even more than in the field, pursuing collaboration after the event needs a fair deal of personal commitment. There will always be tensions between individual and collective research interests, and accordingly, as Dumoulin Kervran (2021, p. 91) writes, 'long-term commitment is not easy to reach, beyond the excitement of being part of an "*historic* high-level conference", as two years is a minimum for this kind of project'. Moreover, he suggests, this means that it is 'risky to give important roles in the project to students and other scholars with very weak positions'. Successful collaboration after the event therefore also depends on success in acquiring research funding, to fund in-person meetings, proofreading and administrative work (Gray et al., 2019, p. 15). Finally and importantly, collaboration after the event means continuing to support each other, providing career guidance to younger members of the team, and nurturing mentoring relations and friendships.

DO: Convey the richness and thickness of the field, while also reconnecting it to its broader context and social science debates; start joint writing projects soon after the event; leave room for individual publication projects; clearly discuss publication strategies and issues of authorship; support early career researchers; nurture long-term relations.

DON'T: Lose the richness of the field or erase your positionality when writing; let the event's momentum go to waste when returning home; lose track of your own career in the collective project.

6 Collaborative Event Ethnography in Action at COP26

As field sites, COPs can be ethnographically observed as circles which extend from the negotiations to the UN Blue Zone, and from there to the wider COP environment within the host city and beyond (Dahan et al., 2009). This metaphor of circles includes both a spatial-physical and a social dimension (Aykut, Pavenstädt, et al., 2022). In spatial terms, movement across the different zones is regulated by material barriers, concrete walls and strict rules of accreditation and access. In social terms, each circle includes different sets of actors and practices and enables different forms of climate politics. Looking at these circles in conjunction allows to develop a fuller picture of the state of global climate politics than focusing exclusively on the negotiations. It also allows to give voice to different sets of actors and discourses. But this shift in focus also necessitates a shift in observation methods. To follow actors, practices and political dynamics across all three spaces in Glasgow, we had to be there and engage in ethnographic observations with a team. Over the two weeks of the Twenty-sixth United Nations Climate Change Conference of the Parties (COP26) in November 2021, eight researchers were physically in Glasgow, the host city. One additional researcher observed remotely, looking at the daily online documentation and watching webcasts of events. We shared information through regular meetings and a common observation matrix. We also adopted a common focus on performances and dramaturgical practices at the conference, and shared an interest in what we understood to be an ongoing transformation of UN Climate Conferences as a political arena. The two-week-long observation across all three circles yielded rich ethnographic material, including countless pictures, fifty-two interviews and hundreds of pages of fieldnotes which we cannot comprehensively cover or synthesise here. Instead, we chose to illustrate the two points on performance and transformation, as well as two core elements of collaborative event ethnography by presenting two carefully chosen vignettes (a more comprehensive account can be found in Aykut, Pavenstädt, et al., 2022). The first is the *principle of many eyes – many minds*, as the presence of many observers and the discussions among them allow researchers to arrive at a richer and more complex picture of a world event. In a first vignette, we thus aim to understand fundamental moments of movement-media interaction at the Glasgow climate conference through observations made with many eyes (Section 6.1). The second is the *principle of repeated observations* within a given global forum, used to identify changes in governance practices that would be invisible to a one-time observer. In our second vignette, we thus look at Glasgow as an event in a sequence of similar events, focusing on transformations over time (Section 6.2).

6.1 Many Eyes on Moments of Movement-Media Interactions

Among our team of ethnographers observing COP26 in Glasgow in November 2021, we had agreed on a common conceptual anchor, namely, conceptualising Glasgow as a *stage* for a wide range of *actors* within the three *circles* of the conference. Two of our team, Max and Simone, had been residents in Glasgow in the past, and neither of them had ever seen the city this neat and tidy (see also Section 3.1). Walking down Buchanan Street from Queen Street Station, Simone remembered hearing the phrase "bright-eyed and bushy-tailed" for the first time when she was a student at Glasgow university many years ago. That a bright-eyed and bushy-tailed city stage was being prepared was reflected in the words of a Glasgow resident our observer Stefan met in a local pub. This resident explained that while they 'care about Glasgow, care about climate', in their view 'this is just bringing PR people and big business into Glasgow. Glasgow has never been so clean. They hired supplementary staff to collect the trash just two weeks before the COP. But we know that everything will go back to normal when the people leave'. This sense that a shiny, but ephemeral frontstage was being set up at the negotiations was shared by the COP26 Coalition, the overarching group that was preparing to mobilise activists of all kinds to accompany the conference with their protest action. Inspired by Goffmanian sociology, the social movement group within our team – Simone, Christopher, Ella and Max – set out to study how individual activists would present themselves in the spotlight provided by the Glasgow stage, and how these presentations drew on various scripts, including scripts in relation to the news media. Applying the principle of many eyes – many minds, we worked in both Edinburgh and Glasgow, and had observers in the Blue and Green UN zones as well as in various activist venues and on the streets. We were commuting from Edinburgh every day to avoid the overpriced Glasgow rates, and adopted the ScotRail service between Glasgow and Edinburgh as our daily meeting space. We had diligently dug through the COP26 Coalition's announcement of protest events prior to our arrival in Scotland, and decided to make a pre-conference Sunday morning march in Edinburgh our first field site. On this Sunday preceding the official opening of COP26, a coalition of protest movements gathered in Glasgow and Edinburgh, announcing their role as *monitoring the official process*: 'WE ARE WATCHING YOU', read the banner at the front of the march in Edinburgh. Similar banners appeared later at other protest events, for example at the climate march on the Global Day of Action for Climate Justice on Saturday 6 November, midway through the conference (Figure 4).

This 'we' was intended as an inclusive one. 'Do you want a flag?' was the welcoming question from the organising Extinction Rebellion stewards to incoming participants, including observers Simone and Ella. On the eve of the arrival of world leaders to a climate COP, protestors know that they will attract media coverage, so the 'We are watching you' implied that the media and thus the world was watching. Balanced reporting is a central norm of journalism and part of a journalistic understanding of objectivity (Westerståhl, 1983), and in the media script for climate conferences, it is social movements that are used to balance coverage of official actors. Simone observed an illustrative example of what Luhmann has called 'a secret alliance between protest movements and the mass media' (1997, p. 855) before the march set off on this crisp autumn morning. Some photojournalists were instrumental in arranging banners and activists and instructed seemingly inexperienced protestors to simultaneously wave their flags for the best possible ensemble presentation (Figure 5a,b).

While the Sunday march was a rather harmonious event with some older ladies among the protestors even blethering away with police officers, Greta

Figure 4 Demonstrations with the slogan WE ARE WATCHING YOU in Glasgow

Figure 5 Media-movement co-production of an ensemble presentation (a, b)

Figure 6 A protest banner at COP15 in Copenhagen (a), posters during COP26 in Glasgow (b)

Thunberg had set the tone of critical movement rhetoric at a pre-conference a few weeks earlier by dismissing the climate negotiations as 'Blah Blah Blah'. While Stefan, a repeat observer of COPs, remembered a picture that he had taken a dozen years ago at the COP15 conference in Copenhagen protest featuring precisely the same dismissive statement (Figure 6a), this time Greta's soundbite seemed to resonate with everyone. It was readily incorporated into the discourse surrounding COP26, and was picked up by demonstrators, representatives, journalists and negotiators as well as on posters throughout Glasgow (Figure 6b).

At the two preceding COPs, Thunberg had spoken prominently inside the conference space. At the Glasgow COP, although she was present in the halls of the official conference, as a speaker she appeared only on the Fring, at protest events. On the first official conference day, Fridays for Future (FfF) held a press conference by the river Clyde with a view of the COP conference area lying across the water, physically illustrating the separation between the two groups. Greta was in attendance but with no visible role. Despite the organisers' reiterated desire to focus the event on the 'most affected people and areas' (termed MAPA) – one organiser explained to the about 200 attendees, including observer Max: 'Just to be clear, Greta won't be speaking' – the attending journalists did not ask questions, but waited for Thunberg to act, to the visible frustration of the other activists. The media's fixation on Thunberg also bothered activists inside the conference venue, as Christopher noted from his Blue Zone observations. One of his interviewees there referred to Thunberg's appearance at this press conference: 'Because it's Greta, there's so much media. Wherever she'll be, there's media, so her message will always matter. I think what's really important is to see what's outside and what's not reported by the media'. Thunberg's iconic status and her consistent attraction of media attention would thus prove a mixed blessing, leaving other groups less able to compete.

The middle weekend of any COP is the time of major protest mobilisation. Halfway through the conference, world leaders have departed, media attention wanes, and a breakthrough in the negotiations is often a distant prospect. And so it was in Glasgow. In the run-up to the Youth Climate Strike, organised by Fridays for Future Scotland with participation from the other national Fridays for Future groups, on the mid-conference Friday, the media communicated an expected number of 25,000 participants, and about that many showed up.

The hilly topography of the Scottish harbour city offered our team a good view of the sea of protesters. Facing in the opposite direction to the march and looking down the street, waves of people swept through after each other: the MAPA bloc, children in school uniforms, a *lot* of trade unions. Most groupings were rather unorganised, lacking visible signs of connection to each other other than groups of students, children with their parents and grandparents, many with hand-painted posters. Among those marching we spotted a couple of polar bears. The bears' message was written on their icy bellies: 'I ♥ nuclear'. From a previous day inside the conference venue, Simone recalled the multiple booths among the country pavilions and in the corporate zone advertising the merits of nuclear power: 'Let the rivers be rivers, and the forests be forests – count on nuclear.' In a spillover from the UN political and corporate spaces, nuclear activists now showed their presence at the Fringe, highly visible in their costumes, yet unchallenged by other protesters. From our German perspective, it was rather surprising that none of the other activists seemed to be bothered by this preference. When we had observed school strikes in Hamburg, anti-nuclear stickers were frequently seen, especially among older demonstrators who had apparently been protesting nuclear energy for many years.

Yet one focus was to be on journalists' perspective, so what self-presentations were the media interested in among the crowd? At the beginning of the rally, Simone noticed a crowd of journalists elevated on a truck at the roadside and watching the march starting and passing by (Figure 7a). She climbed up there as well to observe the march from their angle. At the very front marched a group of Indigenous people or MAPA in their traditional costume, followed by mainly white participants of varied ages. A few minutes passed, during which activists marched and chanted, when suddenly a journalist yelled: 'It's her! There she is!', and there was a sudden burst of clicking cameras. Thunberg was unobtrusively marching among the general crowd of demonstrators, surrounded by a circle of intimates. When she had passed, the clicking and filming abruptly stopped, as the journalists rushed to send the pictures out to the world (Figure 7b). Simone was the last to leave the truck platform, even as many activists continued to pass by.

'Climate justice' was the central slogan of the protests in Glasgow, and the movements used all their stages to bring the representatives of the MAPA to the fore. They were situated at the front of the protest march, and members of Indigenous communities spoke for over half of the two-hour rally at its end. At the very end, Greta spoke for about five minutes. News coverage focused predominantly on Greta. The video of Thunberg's speech at the rally in Glasgow, which she uploaded to her Twitter account, had been viewed more than 1.5 million times as of mid-November 2021. Also on Twitter, fellow activists expressed their displeasure with the traditional news media's obsession with Greta.

Overall, our findings indicate that in the COP situation of high issue attention and overall sympathetic reporting, traditional media selected according to their general news values, and did not transport movement messages unfiltered. In our example, the representatives of the Indigenous MAPA communities were simply ignored. In contrast, social media platforms like Twitter offered activists

Figure 7 Journalists waiting for (a) and covering (b) Greta Thunberg during the Youth Strike at COP26

opportunities to communicate directly with their audiences and, of course, provided spaces for media criticism. The COP26 Coalition also streamed a daily recap on YouTube for those not able to be in Glasgow in person, pointedly called 'Inside Outside'. In-person sessions typically acknowledged who was not there, with many references to this being 'the most exclusionary COP ever'. In conclusion, our many eyes were able to observe a range of movement-media spaces and interactions therein. The experiences we documented among actors on the COP Fringe ranged from co-producing ensemble presentations with journalists and being cast as the favourite critic of the official negotiations to frustration with traditional media gatekeeping and the use of social media to criticise them.

6.2 Tracking the Emergence of the Climate Action Zone

Much like the 'traditional' ethnographer returning regularly to a specific place or people after a period of absence, repeated observations at subsequent conferences within a field of global environmental governance can be a useful strategy to detect changes that would not be perceptible to a first-time observer. These can include more or less gradual shifts in the ways in which these conferences are organised, in governance practices, and the material design of the venue. To illustrate this point, the following vignette draws together snippets of fieldwork conducted with three different teams between 2018 and 2021 at COP24 in Katowice, Poland, COP25 in Madrid, Spain, and COP26 in Glasgow, Scotland, to document the evolution of the climate action space. Although each of these observations followed a much larger research agenda, the vignette focuses exclusively on observations relevant to this subject.

The COP24 in Katowice was the first to feature a so-called Climate Action Hub. When our team first discovered this space in the form of a small, half-open amphitheatre (Figure 8a,b) situated immediately after the entrance and before the plenary rooms, we did not immediately understand its function, nor think much of its relevance. In the words of the Polish COP Presidency,

Figure 8 The Climate Action Hub at COP24 (a, b)

Collaborative Ethnography of Global Environmental Governance 57

the area was to 'provide a central scene that is inclusive, participatory and transparent' and 'a dynamic events-space – suitable for staging digital events, talk-shows, launch and announcement events, film screenings, competition winners, on-the-sofa style discussions'. This appeared to be an intriguing concept for an official space within a global climate conference. However, occupied with other themes, we only sporadically attended events within the hub. These included film screenings, a panel discussion with so-called 'climate leaders' and displays of climate action by businesses, cities and bottom-up initiatives.

The idea of a Climate Action Hub was taken up a year later at COP25 in Madrid. There it drew our attention considerably more. Its repeated presence suggested that rather than being a one-off initiative, this was liable to be a new and lasting feature of the UNFCCC governance space. The space was again designed as a half-open amphitheatre, and was used to provide a stage for presenting transnational initiatives and non-state climate action (Figure 9a,b). In our COP observation report, we noted:

> The tone and style of events at the Action Hub stood in stark contrast not only to the negotiations, but also to other side events. City mayors, investors, and foundations took it in turns to present their climate actions; startups and businesses introduced their latest technological solutions; young activists shared success stories of youth mobilization; and media channels committed to enhancing their coverage of climate issues. The agenda combined TED talk-style and talk show-like formats, as well as interactive events, movie projections, and artistic performances. (Aykut et al., 2020)

The Global Climate Action Awards Ceremony on the Tuesday of the second week of COP25 exemplified the spirit of the space. Announced as 'a moment of celebration ... with inspiring speakers, videos, photography, and a musical performance', it presented the fifteen winners of the 2019 UN Global Climate Action Awards. Awards were given, for instance, to a company that invented a technology to generate energy from ocean waves, an initiative for women's

Figure 9 The Climate Action Hub at COP25 in Madrid (a, b)

empowerment in agriculture, and the Swedish fast food chain MAX Burgers for proposing a 'climate-positive' menu. The event supported and illustrated the narrative of a 'groundswell' of climate action driven by pioneering individuals, local governments and companies. UNFCCC Deputy Executive Secretary Ovais Sarmad stated in his introduction, 'agreements are needed, but these actions give them the meaning, give them the momentum they need'. Adventurer and entrepreneur Bertrand Piccard, who moderated the event, concluded with a similar optimism: 'If we look here, I think we can be optimistic. The whole first row is full with innovators that act now.'

At COP26 in Glasgow, the UK Presidency took this idea of a dedicated climate action space to a completely new level, in terms of size and spatiality, but also media presence and communication efforts. Rechristened Climate Action Zone, the venue was conceived as a huge open space, with several stages and a massive illuminated globe representing the planet Earth hanging in the middle. Several members of our team noted that the latter proved a very popular spot for selfies. It was equally popular as a visual background for international media outlets' live broadcasting from within the COP venue (Figure 10). Overlooking this space on all sides was a sports arena-like architecture, composed of rows of seats stretching high above ground level, from which spectators could attain a synoptic view of events within the Action Zone. The UK COP Presidency described the area as 'a dynamic events-space, where non-Party stakeholders can stage a variety of events, such as talk-shows, special launch events, competition winners announcements, games, interactive activities or digital demonstrations, all of which focus on concrete climate action and provide a voice to the audience'. In our observation notes and team meetings,

Figure 10 The Climate Action Zone in Glasgow

we repeatedly noted that the corporate presence in the climate action space appeared greater than in previous years. It manifested in the type of events taking place in the Climate Action Zone, in the ways in which speakers and audiences were dressed and in the very 'corporate' language used in panel discussions, but also in the continuous displays of the logos of COP26 sponsors on the arena walls (Bloomberg, Microsoft, Google). Events provided ample room for firms and start-ups, city leaders and young entrepreneurs to put their activities on display in multimedia presentations, TED-style talks, and award ceremonies. Finally, despite or because of this corporate presence, the Zone was also a popular stage for protests by Fridays for Future and other climate activists.

Overall our observations show how the spirit of the Paris Agreement, with its focus on momentum building, emotional communication strategies, and a steady stream of pledges and promises (Aykut et al., 2021), gradually found a material embodiment at climate conferences in the form of a newly created climate action space. The growing size and centrality of this space within the COP symbolises a broader reorientation of climate governance. Climate action by 'non-Party stakeholders' forms a core pillar of the architecture of the Paris Agreement, and various processes have been created to reach beyond governments and directly address wider society. But the aim of showcasing private initiatives and positive storytelling sits rather uneasily with the traditional diplomatic language and formalised procedures of international negotiations. The creation of a dedicated climate action space that is situated within the UNFCCC process, but outside the negotiations, represents an attempt to resolve this problem. The Climate Action Zone supports the narrative of a global 'groundswell' of climate action through events like the annual Climate Action Awards Ceremony. The Secretariat facilitated these and other activities by providing specific Global Climate Action badges for representatives of civil society, start-ups, businesses and cities, and by setting up a dedicated Climate Action Unit. New actors are thus encouraged to populate climate governance arenas and exhibit their ideas in public and with media scrutiny.

This demonstrates the value of using ethnographic methods and repeated COP observations to understand changes in global environmental governance. Indeed, neither the Paris Agreement nor subsequent COP decisions are very specific about non-state action (van Asselt et al., 2018, pp. 30–31). COP Presidencies and the Secretariat creatively navigate this area of legal uncertainty by using their organisational prerogatives to organise public events that shed light on specific actors or agenda items, and by designing the material venue and the stage set for public performances. Using emotional communication frames such as alarmist messaging and positive storytelling, they significantly shape the narrative arc of COPs. This attracts media attention and positions the UN process as a necessary part of the solution.

However, as academic observers from European universities, the enthusiastic language, uplifting tone and showy character of many interventions in climate action spaces was also bewildering to us, standing as it did in stark and uncomfortable contrast to the ongoing climate destruction mentioned in other areas of the conference. This dissonance was the subject of regular discussions among members of the group, in which some expressed a view of these events as cynical displays of greenwashing, whereas others mostly stressed the increasing pervasiveness of management language and business culture at UN events. These discussions made it clear to us that the specific dramaturgy and choreography at COPs should not be seen as merely apolitical means to reform the COP process or 'orchestrate' global climate action, but that it was important to recognise its tendency to stabilise the dominant paradigm of liberal environmentalism in climate debates. In a context of growing critiques of capitalism and extractivism within activist spaces, but also well beyond them, the public arena provided by global climate conferences continues to foreground decontextualised 'best practices' and technofixes, and to dissociate spaces for systemic critique and discussions about causes from seemingly pragmatic spaces for and discussions on solutions.

7 Concluding Remarks

Transnational mega-events such as climate and biodiversity COPs appear likely to constitute a lasting feature of earth system governance well into the future. Collaborative event ethnography is thus here to stay. But observing global environmental conferences requires substantial personal, financial and time resources as well as institutional access to the field. Access is gradually improving as more civil society observers are progressively included in UN processes, a development that also provides access for ethnographic participant observation by way of a formal observer status. Still, researchers need to obtain formal accreditation through an eligible institution. Given its resource-intensive and complex nature, the question of when and why this research approach should be chosen is an important one. When does it make sense to 'be there' as a team? And when, to the contrary, are document analyses, online observations and interviews after the event enough?

These questions cannot be answered once and for all. As with any methodology, the pros and cons of collaborative event ethnography are context-dependent. However, when the approach fits the research object and is well implemented, collaborative event ethnography has clear benefits. As far as we can see, broad research questions with some degree of abstraction and that are difficult to study empirically as an individual researcher, lend themselves particularly well to being studied through collaborative event ethnography. With its many eyes, many minds approach, it provides a unique vantage point to understand meaning-making processes in global governance spaces. These processes involve the plurality of actors within these spaces, as well as the researcher, who co-constitute this meaning. Being there, then, facilitates access to potential interviewees and otherwise inaccessible informants, and thus helps to circumvent some of the common obstacles to 'studying up'. It provides insights into the bodily experience of attending a global environmental conference, and of being exposed to its emotional impact and dramaturgical arc, occasions to observe stage-setting practices, and glimpses into the backstage of the shiny world of global governance. Perhaps most importantly, good ethnography can give the reader a sense of why the site in itself is puzzling, its peculiarities, and how it affects the practice and the politics of global governance. The ultimate benefit of this approach is its contribution to broader research on IR and global environmental politics by going beyond document analysis and interviews. Collaborative event ethnography allows for experiencing and reflecting, as well as analysing, the temporal and spatial dynamics of world events which run for several weeks and span a multiplicity of spaces, venues and arenas. The dramaturgical focus allows for studying front stage as well as backstage dynamics as they constitute one another as well as the interconnections between scripts, stages and settings, all taking place within complex temporal dynamics. It makes a difference whether one reads about 40,000 participants or whether one experiences them flocking in and out of the conference hall, being dispersed throughout multiple conference spaces and fluctuating between the circles.

Going on trips into the field with a social science team – a practice that is more common in the natural sciences – is also a highly enriching and rewarding experience. The formation of a caring and supportive team, in both intellectual and social terms, provides mentoring opportunities and important avenues for professional development. The flexibility and adaptability of the methodology to different field sites enable the development of wide-ranging and nuanced understandings of what is going on. Still, team dynamics should be closely monitored with empathy and sensitivity, and questions of authorship and ownership of intellectual property should be discussed transparently beforehand, throughout and following the fieldwork. After all, with its colonial legacy and due to its time-consuming nature, ethnography is known to be a highly unequal research practice. Ethnographers of global environmental governance need to reflect on inequalities along gender, class, racial and colonial lines, and handle power asymmetries reflexively. This includes building non-extractive research collaborations with informants and showing solidarity with colleagues from the Global South as well as younger scholars without adequate research funding.

To close, let us emphasise the integrative potential of collaborative event ethnography. While the integration and negotiation of different unique field experiences can be challenging – especially across different theoretical and methodological orientations – preparatory seminars, regular exchanges during the field trip and writing retreats after the event can help to forge a common perspective or develop a shared set of ideas around an anchoring concept. Collaborative event ethnography thus brings innovative angles to research on IR, social movements and earth system governance. For example, research on international organisations and environmental regimes still tends to be done separately from research on transnational governance networks involving NGOs, firms, science advisers and city governments, or research on protest movements and transnational mobilisations. Collaborative event ethnography, as a methodological approach and collaborative research practice, has the potential to integrate these fields to more comprehensively study the dynamics of earth system governance. It allows us to assemble a bigger picture through collaboration and collective research, bringing together many eyes and many minds, not only at and around a single event such as COP26, but also through repeated observations of the same kind of events that track their changes through time, and through comparison with other global events that aim to tackle comparable problems of planetary concern.

References

Abidin, C. & de Seta, G. (2020). Doing Digital Ethnography: Private Messages from the Field. *Journal of Digital Social Research*, *2*(1), 1–19.

Ariffin, Y., Coicaud, J.-M. & Popovski, V. (2016). *Emotions in International Politics: Beyond Mainstream International Relations*. Cambridge University Press.

Atkinson, P., Coffey, A., Delamont, S., Lofland, J. & Lofland, L. (Eds.). (2011). *Handbook of Ethnography*. SAGE.

Aubertin, C. (2015). L'art de la diplomatie en temps de crise: Débats sur le développement durable au Brésil. In J. Foyer (Ed.), *Regards croisés sur Rio+20: La modernisation écologique à l'épreuve* (pp. 117–136). CNRS Éditions. https://doi.org/10.4000/books.editionscnrs.26304.

Augé, M. (1995). *Non-Places: Introduction to an Anthropology of Supermodernity*. Verso.

Aykut, S. C. (2017). Governing through Verbs: The Practice of Negotiating and the Making of a New Mode of Governance. In S. C. Aykut, J. Foyer & E. Morena (Eds.), *Globalising the Climate COP21 and the Climatisation of Global Debates* (pp. 18–38). Routledge.

Aykut, S. C. (2020). Global by Nature? Three Dynamics in the Making of "Global Climate Change". In E. Neveu & M. Surdez (Eds.), *Globalizing Issues: How Claims, Frames, and Problems Cross Borders* (pp. 277–299). Springer International. https://doi.org/10.1007/978-3-030-52044-1_13.

Aykut, S. C., d'Amico, E., Klenke, J. & Schenuit, F. (2020). The Accountant, the Animator, and the Admonisher: Global Climate Governance in Transition. Report from the COP25 climate summit in Madrid. *CSS Working Paper, No 1*. https://doi.org/10.25592/css-wp-001.

Aykut, S. C. & Dahan, A. (2011). Le régime climatique avant et après Copenhague: Sciences, politiques et l'objectif des deux degrés ». *Natures Sciences Sociétés*, *19*(2), 144–157.

Aykut, S. C., Foyer, J. & Morena, E. (Eds.). (2017). *Globalising the Climate: COP21 and the Climatisation of Global Debates*. Routledge.

Aykut, S. C. & Maertens, L. (2021). The Climatization of Global Politics: Introduction to the Special Issue. *International Politics*, *58*(4), 501–518. https://doi.org/10.1057/s41311-021-00325-0.

Aykut, S. C., Morena, E. & Foyer, J. (2021). 'Incantatory' Governance: Global Climate Politics' Performative Turn and Its Wider Significance for Global

Politics. *International Politics*, *58*(4), 519–540. https://doi.org/10.1057/s41311-020-00250-8.

Aykut, S. C., Pavenstädt, C. N., Datchoua-Tirvaudey, A., et al. (2022). *Circles of Global Climate Governance: Power, Performance and Contestation at the UN Climate Conference COP26 in Glasgow*. https://doi.org/10.25592/CSS-WP-004.

Aykut, S. C., Schenuit, F., Klenke, J. & d'Amico, E. (2022). It's a Performance, Not an Orchestra! Rethinking Soft Coordination in Global Climate Governance. *Global Environmental Politics*, *22*(4), 1–24. https://doi.org/10.1162/glep_a_00675.

Baillot, H., Bergamaschi, I. & Iori, R. (2015). Division of Labor and Partnerships in Transnational Social Movements. Observations of North-South and South-South Interactions at the World Social Forum. In J. Siméant, I. Sommier, & M.-E. Pommerolle (Eds.), *Observing Protest from a Place* (pp. 115–136). Amsterdam University Press. https://doi.org/10.1515/9789048525805-007.

Becker, H. S., Geer, B., Hughes, E. C. & Strauss, A. L. (1977). *Boys in White: Student Culture in Medical School*. Transaction Books.

Bellier, I. (2012). Les peuples autochtones aux Nations unies: Un nouvel acteur dans la fabrique des normes internationales. *Critique Internationale*, *1*, 61–80.

Benabou, S., Moussu, N. & Müller, B. (2017). The Business Voice at COP21: The Quandaries of a Global Political Ambition. In S. C. Aykut, J. Foyer & E. Morena (Eds.), *Globalising the Climate* (pp. 57–74). Routledge.

Blühdorn, I. (2007). Sustaining the Unsustainable: Symbolic Politics and the Politics of Simulation. *Environmental Politics*, *16*(2), 251–275. https://doi.org/10.1080/09644010701211759.

Boden, D. & Molotch, H. L. (1994). The Compulsion of Proximity. In R. Friedland & D. Boden (Eds.), *NowHere: Space, Time, and Modernity* (pp. 257–286). University of California Press.

Breidenstein, G. & Hirschauer, S. (2002). Weder 'Ethno' noch 'Graphie'. Anmerkungen zu Hubert Knoblauchs Beitrag 'Fokussierte Ethnographie'. *Sozialer Sinn*, *1*, 125–129.

Brosius, J. P. & Campbell, L. M. (2010). Collaborative Event Ethnography: Conservation and Development Trade-Offs at the Fourth World Conservation Congress. *Conservation and Society*, *8*(4), 245–255. https://doi.org/10.4103/0972-4923.78141.

Brown, H., Reed, A. & Yarrow, T. (Eds.). (2017). *Meetings: Ethnographies of Organizational Process, Bureaucracy, and Assembly*. John Wiley & Sons.

Bueger, C., & Gadinger, F. (2014). Die Formalisierung der Informalität: Praxistheoretische Überlegungen. In S. Bröchler & T. Grunden (Eds.), *Informelle Politik* (pp. 81–98). Springer.

References

Buford May, R. A. & Pattillo-McCoy, M. (2000). Do You See What I See? Examining a Collaborative Ethnography. *Qualitative Inquiry*, 6(1), 65–87. https://doi.org/10.1177/107780040000600105.

Burawoy, M. (2000). Introduction: Reaching for the Global. In M. Burawoy, J. A. Blum, S. George, et al. (Eds.), *Global Ethnography* (pp. 1–41). University of California Press.

Burawoy, M. (2001). Manufacturing the Global. *Ethnography*, 2(2), 147–159. https://doi.org/10.1177/1466138101002002001.

Burawoy, M. (2003). Revisits: An Outline of a Theory of Reflexive Ethnography. *American Sociological Review*, 68(5), 645–679. https://doi.org/10.2307/1519757.

Büscher, B. (2014). Collaborative Event Ethnography: Between Structural Power and Empirical Nuance? *Global Environmental Politics*, 14(3), 132–138. https://doi.org/10.1162/GLEP_a_00243.

Campbell, L. M., Corson, C., Gray, N. J., MacDonald, K. I. & Brosius, J. P. (2014). Studying Global Environmental Meetings to Understand Global Environmental Governance: Collaborative Event Ethnography at the Tenth Conference of the Parties to the Convention on Biological Diversity. *Global Environmental Politics*, 14(3), 1–20. https://doi.org/10.1162/GLEP_e_00236.

Campbell, L. M., Gray, N. J., Fairbanks, L. W., Silver, J. J. & Gruby, R. L. (2013). Oceans at Rio+20. *Conservation Letters*, 6(6), 439–447. https://doi.org/10.1111/conl.12035.

Campbell, L. M., Hagerman, S. & Gray, N. J. (2014). Producing Targets for Conservation: Science and Politics at the Tenth Conference of the Parties to the Convention on Biological Diversity. *Global Environmental Politics*, 14(3), 41–63. https://doi.org/10.1162/GLEP_a_00238.

Chaffee, L. (1993). Dramaturgical Politics: The Culture and Ritual of Demonstrations in Argentina. *Media, Culture & Society*, 15(1), 113–135. https://doi.org/10.1177/016344393015001009.

Chartier, D. & Le Crosnier, H. (2015). Acter la fin d'un monde pour activer l'alternative: ONG et acteurs de l'altermondialisme à Rio+20. In J. Foyer (Ed.), *Regards croisés sur Rio+20: La modernisation écologique à l'épreuve* (pp. 281–304). CNRS Éditions. https://doi.org/10.4000/books.editionscnrs.26352.

Chasek, P. (2021). Is It the End of the COP as We Know It? An Analysis of the First Year of Virtual Meetings in the UN Environment and Sustainable Development Arena. *International Negotiation*, 28(1), 1–32. https://doi.org/10.1163/15718069-bja10047.

Clayton, S. (2020). Climate Anxiety: Psychological Responses to Climate Change. *Journal of Anxiety Disorders*, 74, 1–7.

Clément, M. & Sangar, E. (2018). *Researching Emotions in International Relations: Methodological Perspectives on the Emotional Turn.* Palgrave Macmillan.

Clerke, T. & Hopwood, N. (2014). *Doing Ethnography in Teams.* Springer International. https://doi.org/10.1007/978-3-319-05618-0.

Clifford, J. & Marcus, G. E. (Eds.). (1986). *Writing Culture: The Poetics and Politics of Ethnography.* University of California Press.

Coleman, S. & von Hellerman, P. (Eds.). (2011). *Multi-sited Ethnography: Problems and Possibilities in the Translocation of Research Methods.* Routledge.

Collins, H., Leonard-Clarke, W. & Mason-Wilkes, W. (2023). Scientific Conferences, Socialization, and the Covid-19 Pandemic: A Conceptual and Empirical Enquiry. *Social Studies of Science, 53*(3), 379–401. https://doi.org/10.1177/03063127221138521.

Collins, H. & Pinch, T. (1982). *Frames of Meaning.* Routledge.

Corson, C., Campbell, L. M. & MacDonald, K. I. (2014). Capturing the Personal in Politics: Ethnographies of Global Environmental Governance. *Global Environmental Politics, 14*(3), 21–40.

Corson, C., Campbell, L. M., Wilshusen, P. & Gray, N. J. (2019). Assembling Global Conservation Governance. *Geoforum, 103,* 56–65. https://doi.org/10.1016/j.geoforum.2019.03.012.

Crawford, N. C. (2014). Institutionalizing Passion in World Politics: Fear and Empathy. *International Theory, 6*(3), 535–557.

Crow, G. M., Levine, L. & Nager, N. (1992). Are Three Heads Better Than One? Reflections on Doing Collaborative Interdisciplinary Research. *American Educational Research Journal, 29*(4), 737–753. https://doi.org/10.3102/00028312029004737.

Dahan, A., Aykut, S., Buffet, C. & Viard-Crétat, A. (2010). Les Leçons politiques de Copenhague: Faut-il repenser le régime climatique. *Koyré Climate Series, 2,* 1–45.

Dahan, A., Aykut, S. C., Guillemot, H. & Korczak, A. (2009). Les arènes climatiques: Forums du futur ou foires aux palabres. La conférence de Poznan. *Koyré Climate Series, 1,* 1–45.

Dahan-Dalmedico, A. (2008). Climate Expertise: Between Scientific Credibility and Geopolitical Imperatives. *Interdisciplinary Science Reviews, 33,* 71–81. https://doi.org/10.1179/030801808X259961.

de Moor, J. (2021). Alternative Globalities? Climatization Processes and the Climate Movement beyond COPs. *International Politics, 58,* 582–599. https://doi.org/10.1057/s41311-020-00222-y.

de Moor, J., Morena, E. & Comby, J.-B. (2017). The Ins and Outs of Climate Movement Activism at COP21 In S. C. Aykut, J. Foyer & E. Morena, (Eds.),

Globalising the Climate. COP21 and the Climatisation of Global Debates (pp. 75–94). Routledge.

Death, C. (2010). Counter-Conducts: A Foucauldian Analytics of Protest. *Social Movement Studies*, 9(3), 235–251. https://doi.org/10.1080/14742837.2010.493655.

Death, C. (2011). Summit Theatre: Exemplary Governmentality and Environmental Diplomacy in Johannesburg and Copenhagen. *Environmental Politics*, 20(1), 1–19. https://doi.org/10.1080/09644016.2011.538161.

Demeulenaere, E. & Castro, M. (2015). Modèles de verdissement de l'agriculture et acteurs en compétition à Rio+20. In J. Foyer (Ed.), *Regards croisés sur Rio+20: La modernisation écologique à l'épreuve* (pp. 185–211). CNRS Éditions. https://doi.org/10.4000/books.editionscnrs.26325.

Doolittle, A. A. (2010). The Politics of Indigeneity: Indigenous Strategies for Inclusion in Climate Change Negotiations. *Conservation and Society*, 8(4), 286–291.

Duffy, R. (2014). What Does Collaborative Event Ethnography Tell Us about Global Environmental Governance? *Global Environmental Politics*, 14(3), 125–131. https://doi.org/10.1162/GLEP_a_00242.

Dumoulin Kervran, D. (2021). Collaborative Event Ethnography as a Strategy for Analyzing Policy Transfers and Global Summits. In O. Porto de Oliveira (Ed.), *Handbook of Policy Transfer, Diffusion and Circulation* (pp. 80–99). Edward Elgar. https://doi.org/10.4337/9781789905601.

Dumoulin Kervran, D. (2015). La grandeur de Rio+20: Formats et enrôlements de la société politique globale. In J. Foyer (Ed.), *Regards croisés sur Rio+20: La modernisation écologique à l'épreuve* (pp. 51–88). CNRS Éditions. https://doi.org/10.4000/books.editionscnrs.26289.

Dunn, D. H. (1996). What Is Summitry? In D. H. Dunn (Ed.), *Diplomacy at the Highest Level: The Evolution of International Summitry* (pp. 3–22). Palgrave Macmillan. https://doi.org/10.1007/978-1-349-24915-2_1.

Edelman, M. J. (1985). *The Symbolic Uses of Politics*. University of Illinois Press.

Ellis, C. (1995). Emotional and Ethical Quagmires in Returning to the Field. *Journal of Contemporary Ethnography*, 24(1), 68–98. https://doi.org/10.1177/089124195024001003.

Emerson, R. M., Fretz, R. I. & Shaw, L. L. (2011). *Writing Ethnographic Fieldnotes*. The University of Chicago Press.

Erickson, K. & Stull, D. (1998). *Doing Team Ethnography: Warnings and Advice*. SAGE.

Evens, T. & Handelman, D. (Eds.). (2006). *The Manchester School: Practice and Ethnographic Praxis in Anthropology*. Berghahn Books.

Falzon, M.-A. (Ed.). (2009). *Multi-sited Ethnography: Theory, Praxis and Locality in Contemporary Research*. Routledge. https://doi.org/10.4324/9781315596389.

Feldman, G. (2011). If Ethnography Is More Than Participant-Observation, Then Relations Are More Than Connections: The Case for Nonlocal Ethnography in a World of Apparatuses. *Anthropological Theory, 11*(4), 375–395. https://doi.org/10.1177/1463499611429904.

Fenno, R. F. (1990). *Watching Politicians: Essays on Participant Observation*. Institute of Governmental Studies, University of California at Berkeley.

Ferguson, R.-H. (2017). Offline 'Stranger' and Online Lurker: Methods for an Ethnography of Illicit Transactions on the Darknet. *Qualitative Research, 17*(6), 683–698.

Fine, G. A. (Ed.). (1995). *A Second Chicago School?: The Development of a Postwar American Sociology*. University of Chicago Press.

Fine, G. A. & Hancock, B. H. (2017). The New Ethnographer at Work. *Qualitative Research, 17*(2), 260–268. https://doi.org/10.1177/1468794116656725.

Foucault, M. (1980). *Power/Knowledge: Selected Interviews and Other Writings, 1972–1977*. Pantheon Books.

Foyer, J. (2015a). Écoverio: Contribuer à la compréhension des évènements internationaux et à une méthodologie collaborative. In J. Foyer (Ed.), *Regards croisés sur Rio+20: La modernisation écologique à l'épreuve* (pp. 29–47). CNRS Éditions. https://doi.org/10.4000/books.editionscnrs.26283.

Foyer, J. (2015b). Introduction: La modernisation écologique à l'épreuve de Rio +20. In J. Foyer (Ed.), *Regards croisés sur Rio+20, la modernisation écologique à l'épreuve* (pp. 11–28). CNRS Éditions.

Foyer, J. (Ed.). (2015c). *Regards croisés sur Rio+ 20: La modernisation écologique à l'épreuve*. CNRS Éditions.

Foyer, J., Aykut, S. C. & Morena, E. (2017). Introduction: COP21 and the 'Climatisation' of Global Debates. In S. C. Aykut, J. Foyer & E. Morena (Eds.), *Globalising the Climate. COP21 and the Climatisation of Global Debates* (pp. 1–17). Routledge.

Foyer, J. & Dumoulin Kervran, D. (2017). Objectifying Traditional Knowledge, Re-enchanting the Struggle against Climate Change. In S. C. Aykut, J. Foyer & E. Morena (Eds.), *Globalising the Climate* (pp. 153–172). Routledge.

Foyer, J. & Morena, É. (2015). Une recherche collaborative pour analyser la conférence Paris Climat 2015: Le projet ClimaCOP. *Natures Sciences Sociétés, 23*(3), 275–279. https://doi.org/10.1051/nss/2015051.

Galbraith, J. (2015). Deadlines as Behavior in Diplomacy and International Law. *U of Penn Law School, Public Law Research Paper*, 15–16. https://papers.ssrn.com/sol3/papers.cfm?abstract_id=2623649.

Garcia, A. C., Standlee, A. I., Bechkoff, J. & Cui, Y. (2009). Ethnographic Approaches to the Internet and Computer-Mediated Communication. *Journal of Contemporary Ethnography*, 38(1), 52–84. https://doi.org/10.1177/0891241607310839.

Garfinkel, H. (1967). *Studies in Ethnomethodology*. Prentice Hall.

Geertz, C. (1973). Thick Description: Towards an Interpretative Theory of Culture. In C. Geertz (Ed.), *The Interpretation of Cultures* (pp. 310–323). Basic Books.

Gerstl-Pepin, C. I. & Gunzenhauser, M. G. (2002). Collaborative Team Ethnography and the Paradoxes of Interpretation. *International Journal of Qualitative Studies in Education*, 15(2), 137–154. https://doi.org/10.1080/09518390110111884.

Goffman, E. (1959). *The Presentation of Self in Everyday Life*. Anchor.

Goffman, E. (1963). *Behavior in Public Places*. The Free Press.

Goffman, E. (1989). On Fieldwork. *Journal of Contemporary Ethnography*, 18(2), 123–132.

González-Hidalgo, M. & Zografos, C. (2020). Emotions, Power, and Environmental Conflict: Expanding the 'Emotional Turn' in Political Ecology. *Progress in Human Geography*, 44(2), 235–255.

González-Santos, S. & Dimond, R. (2015). Medical and Scientific Conferences as Sites of Sociological Interest: A Review of the Field. *Sociology Compass*, 9(3), 235–245. https://doi.org/10.1111/soc4.12250.

Gouldner, A. W. (1968). The Sociologist as Partisan: Sociology and the Welfare State. *The American Sociologist*, 3(2), 103–116.

Gray, N. J. (2010). Sea Change: Exploring the International Effort to Promote Marine Protected Areas. *Conservation and Society*, 8(4), 331–338.

Gray, N. J., Corson, C., Campbell, L. M., et al. (2019). Doing Strong Collaborative Fieldwork in Human Geography. *Geographical Review*. https://doi.org/10.1111/gere.12352.

Gray, N. J., Gruby, R. L. & Campbell, L. M. (2014). Boundary Objects and Global Consensus: Scalar Narratives of Marine Conservation in the Convention on Biological Diversity. *Global Environmental Politics*, 14(3), 64–83. https://doi.org/10.1162/GLEP_a_00239.

Guillemot, H. (2017). The Necessary and Inaccessible 1.5 C Objective: A Turning Point in the Relations between Climate Science and Politics? In S. C. Aykut, J. Foyer & E. Morena (Eds.), *Globalising the Climate* (pp. 39–56). Routledge.

Gusfield, J. R. (1984). *The Culture of Public Problems: Drinking-Driving and the Symbolic Order*. University of Chicago Press.

Gusterson, H. (1997). Studying Up Revisited. *Political and Legal Anthropology Review*, *20*(1), 114–119.

Hagerman, S., Satterfield, T. & Dowlatabadi, H. (2010). Climate Change Impacts, Conservation and Protected Values: Understanding Promotion, Ambivalence and Resistance to Policy Change at the World Conservation Congress. *Conservation and Society*, *8*(4), 298–311.

Hajer, M. A. (1997). *The Politics of Environmental Discourse: Ecological Modernization and the Policy Process*. Oxford University Press. https://doi.org/10.1093/019829333X.001.0001.

Hajer, M. A. (2009). *Authoritative Governance: Policy-Making in the Age of Mediatization*. Oxford University Press.

Hall, T. H. (2015). *Emotional Diplomacy: Official Emotion on the International Stage*. Cornell University Press.

Hammersley, M. & Atkinson, P. (2019). *Ethnography: Principles in Practice*. Routledge.

Hannerz, U. (2003). Being There ... and There ... and There! Reflections on Multi-site Ethnography. *Ethnography*, *4*(2), 201–216.

Hawkes, N. (1972). Stockholm: Politicking, Confusion, but Some Agreements Reached. *Science*, *176*(4041), 1308–1310.

Heintz, B. (2014). Die Unverzichtbarkeit von Anwesenheit. Zur weltgesellschaftlichen Bedeutung globaler Interaktionssysteme / Personal Encounters. The Indispensability of Face-to-Face Interaction at the Global Level. *Zeitschrift Für Soziologie, Sonderheft "Interaktion, Organisationn und Gesellschaft"*, 229–250. https://doi.org/10.1515/9783110509243-013.

Herrera, W., Judell, A. & Paule, C. (2015). Making Waste (in)visible at the Dakar World Social Forum: A Goffmanian Perspective on a Transnational Alter-Global Gathering. In J. Siméant, M.-E. Pommerolle & I. Sommier (Eds.), *Observing Protest from a Place* (pp. 137–156). Amsterdam University Press. www.jstor.org/stable/j.ctt16vj27n.10.

Hertz, R. & Imber, J. B. (Eds.). (1995). *Studying Elites Using Qualitative Methods*. SAGE.

Higginbottom, G., Pillay, J. & Boadu, N. (2013). Guidance on Performing Focused Ethnographies with an Emphasis on Healthcare Research. *The Qualitative Report*, *18*(9), 1–17. https://doi.org/10.46743/2160-3715/2013.1550.

Hine, C. (2000). *Virtual Ethnography*. SAGE.

Hitchner, S. (2010). Heart of Borneo as a 'Jalan Tikus': Exploring the Links between Indigenous Rights, Extractive and Exploitative Industries, and Conservation at the World Conservation Congress 2008. *Conservation and Society*, *8*(4), 320–330. https://doi.org/10.4103/0972-4923.78148.

Honer, A. & Hitzler, R. (2015). Life-World-Analytical Ethnography: A Phenomenology-Based Research Approach. *Journal of Contemporary Ethnography*, *44*(5), 544–562. https://doi.org/10.1177/0891241615588589.

Hoppe, R. (2011). *The Governance of Problems: Puzzling, Powering and Participation*. Policy Press. https://doi.org/10.1332/policypress/9781847429629.001.0001.

Hsu, W. F. (2014). Digital Ethnography toward Augmented Empiricism: A New Methodological Framework. *Journal of Digital Humanities*, *3*(1), 3–1.

Hughes, H. & Vadrot, A. B. (2019). Weighting the World: IPBES and the Struggle over Biocultural Diversity. *Global Environmental Politics*, *19*(2), 14–37.

Hutchison, E. & Bleiker, R. (2014). Theorizing Emotions in World Politics. *International Theory*, *6*(3), 491–514.

Hyndman, J. (2001). The Field as Here and Now, Not There and Then. *Geographical Review*, *91*(1/2), 262–272. https://doi.org/10.2307/3250827.

Ibrahim, Y., Rödder, S. & Schnegg, M. (2024). World Organisations, World Events and World Objects: How Science, Politics, and the Mass Media Co-produce Climate Futures. *Globalizations*, *21* (0), 70–87. https://doi.org/10.1080/14747731.2023.2215409.

Jackson, P. T. (2008). Can Ethnographic Techniques Tell Us Distinctive Things about World Politics? *International Political Sociology*, *2*(1), 91–93. https://doi.org/10.1111/j.1749-5687.2007.00035_5.x.

Jarzabkowski, P., Bednarek, R. & Cabantous, L. (2015). Conducting Global Team-Based Ethnography: Methodological Challenges and Practical Methods. *Human Relations*, *68*(1), 3–33. https://doi.org/10.1177/0018726714535449.

Juris, J. S. (2008). Performing Politics: Image, Embodiment, and Affective Solidarity during Anti-corporate Globalization Protests. *Ethnography*, *9*(1), 61–97. https://doi.org/10.1177/1466138108088949.

Knoblauch, H. (2005). Focused Ethnography. *Forum Qualitative Sozialforschung / Forum: Qualitative Social Research*, *6*(3), 1–14, Article 3. https://doi.org/10.17169/fqs-6.3.20.

Knorr-Cetina, K. (1981). *The Manufacture of Knowledge: An Essay on the Constructivist and Contextual Nature of Science*. Pergamon Press.

Koschut, S. (2018). The Power of (emotion) Words: On the Importance of Emotions for Social Constructivist Discourse Analysis in IR. *Journal of International Relations and Development*, *21*, 495–522.

Koschut, S. (2020). *The Power of Emotions in World Politics*. Routledge.

Lascoumes, P. (1994). *L'Éco-pouvoir: Environnements et politiques*. La découverte.

Lassiter, L. E. (2005). *The Chicago Guide to Collaborative Ethnography*. University of Chicago Press.

Latour, B. & Woolgar, S. (1986). *Laboratory Life: The Construction of Scientific Facts*. Princeton University Press.

Laux, H. (2017). Clockwork Society: Die Weltklimakonferenz von Paris als Arena gesellschaftlicher Synchronisation. In A. Henkel, H. Laux & F. Anicker (Eds.), *Raum und Zeit: Soziologische Beobachtungen zur gesellschaftlichen Raumzeit* (pp. 246–279). Beltz Juventa.

Little, P. E. (1995). Ritual, Power and Ethnography at the Rio Earth Summit. *Critique of Anthropology*, *15*(3), 265–288. https://doi.org/10.1177/0308275X9501500303.

Lövbrand, E., Hjerpe, M. & Linnér, B.-O. (2017). Making Climate Governance Global: How UN Climate Summitry Comes to Matter in a Complex Climate Regime. *Environmental Politics*, *26*(4), 580–599. https://doi.org/10.1080/09644016.2017.1319019.

Low, S., Taplin, D. & Lamb, M. (2005). Battery Park City: An Ethnographic Field Study of the Community Impact of 9/11. *Urban Affairs Review*, *40*(5), 655–682. https://doi.org/10.1177/1078087404272304.

Luhmann, N. (1997). *Die Gesellschaft der Gesellschaft* (Vol. 2). Suhrkamp.

MacDonald, K. I. (2010). Business, Biodiversity and New 'Fields' of Conservation: The World Conservation Congress and the Renegotiation of Organisational Order. *Conservation and Society*, *8*(4), 256–275.

MacDonald, K. I. & Corson, C. (2012). 'TEEB Begins Now': A Virtual Moment in the Production of Natural Capital. *Development and Change*, *43*(1), 159–184. https://doi.org/10.1111/j.1467-7660.2012.01753.x.

MacKay, J. & Levin, J. (2015). Hanging Out in International Politics: Two Kinds of Explanatory Political Ethnography for IR. *International Studies Review*, *17*(2), 163–188.

Maclin, E. M. (2010). The 2009 UN Climate Talks: Alternate Media and Participation from Anthropologists. *American Anthropologist*, *112*(3), 464–466. https://doi.org/10.1111/j.1548-1433.2010.01257.x.

Maclin, E. M. & Bello, J. L. D. (2010). Setting the Stage for Biofuels: Policy Texts, Community of Practice, and Institutional Ambiguity at the Fourth World Conservation Congress. *Conservation and Society*, *8*(4), 312–319.

Malinowski, B. (2004). *Argonauts of the Western Pacific: An Account of Native Enterprise and Adventure in the Archipelagoes of Melanesian New Guinea*. Taylor & Francis. (Original work published 1922).

Marcus, G. E. (1983). *Elites, Ethnographic Issues*. University of New Mexico Press.

Marcus, G. E. (1995). Ethnography in/of the World System: The Emergence of Multi-sited Ethnography. *Annual Review of Anthropology*, *24*(1), 95–117. https://doi.org/10.1146/annurev.an.24.100195.000523.

Marion Suiseeya, K. R. & Zanotti, L. (2019). Making Influence Visible: Innovating Ethnography at the Paris Climate Summit. *Global Environmental Politics*, *19*(2), 38–60. https://doi.org/10.1162/glep_a_00507.

Markham, A. N. (2013). Fieldwork in Social Media: What Would Malinowski Do? *Qualitative Communication Research*, *2*(4), 434–446. https://doi.org/10.1525/qcr.2013.2.4.434.

Mauthner, N. S. & Doucet, A. (2008). 'Knowledge Once Divided Can Be Hard to Put Together Again': An Epistemological Critique of Collaborative and Team-Based Research Practices. *Sociology*, *42*(5), 971–985. https://doi.org/10.1177/0038038508094574.

Mazie, S. V. & Woods, P. J. (2003). Prayer, Contentious Politics, and the Women of the Wall: The Benefits of Collaboration in Participant Observation at Intense, Multifocal Events. *Field Methods*, *15*(1), 25–50. https://doi.org/10.1177/1525822X02239571.

Meek, D. (2015). Counter-Summitry: La Via Campesina, the People's Summit, and Rio+20. *Global Environmental Politics*, *15*(2), 11–18.

Merry, S. E. (2006). New Legal Realism and the Ethnography of Transnational Law. *Law & Social Inquiry*, *31*(4), 975–995.

Meyer, J. W. (2010). World Society, Institutional Theories, and the Actor. *Annual Review of Sociology*, *36*(1), 1–20. https://doi.org/10.1146/annurev.soc.012809.102506.

Millen, D. R. (2000). Rapid Ethnography: Time Deepening Strategies for HCI Field Research. *Proceedings of the 3rd Conference on Designing Interactive Systems: Processes, Practices, Methods, and Techniques*, 280–286. https://doi.org/10.1145/347642.347763.

Miller, D. & Slater, D. (2001). *The Internet: An Ethnographic Approach*. Berg.

Monfreda, C. (2010). Setting the Stage for New Global Knowledge: Science, Economics, and Indigenous Knowledge in 'The Economics of Ecosystems and Biodiversity' at the Fourth World Conservation Congress. *Conservation and Society*, *8*(4), 276–285.

Montsion, J. M. (2018). Ethnography and International Relations: Situating Recent Trends, Debates and Limitations from an Interdisciplinary Perspective. *The Journal of Chinese Sociology*, *5*(1), 1–21. https://doi.org/10.1186/s40711-018-0079-4.

Morena, E. (2021). The Climate Brokers: Philanthropy and the Shaping of a 'US-Compatible' International Climate Regime. *International Politics*, *58*(4), 541–562. https://doi.org/10.1057/s41311-020-00249-1.

Mountz, A., Bonds, A., Mansfield, B., et al. (2015). For Slow Scholarship: A Feminist Politics of Resistance through Collective Action in the Neoliberal

University. *ACME: An International Journal for Critical Geographies*, *14*(4), 1235–1259.

Muecke, M. A. (1994). On the Evaluation of Ethnographies. In J. M. Morse (Ed.), *Critical Issues in Qualitative Research Methods* (pp. 187–209). SAGE.

Müller, B. (2012). Comment rendre le monde gouvernable sans le gouverner: Les organisations internationales analysées par les anthropologues. *Critique Internationale*, *54*, 9–18.

Müller, B. (Ed.). (2013). *The Gloss of Harmony: The Politics of Policy-Making in Multilateral Organisations*. Pluto Press.

Müller, B. & Cloiseau, G. (2015). The Real Dirt on Responsible Agricultural Investments at Rio+ 20: Multilateralism versus Corporate Self-Regulation. *Law & Society Review*, *49*(1), 39–67.

Nader, L. (1972). Up the Anthropologist — Perspectives Gained from Studying Up. In D. Hymes (Ed.), *Reinventing Anthropology* (pp. 284–311). Pantheon Books.

Neumann, I. B. (2007). "A Speech That the Entire Ministry May Stand for," or: Why Diplomats Never Produce Anything New. *International Political Sociology*, *1*(2), 183–200. https://doi.org/10.1111/j.1749-5687.2007.00012.x.

Obergassel, W., Bauer, S., Hermwille, L., et al. (2022). From Regime-Building to Implementation: Harnessing the UN Climate Conferences to Drive Climate Action. *Wiley Interdisciplinary Reviews – Climate Change*, *13*(6), 1–12. https://doi.org/10.1002/wcc.797.

O'Neill, K. & Haas, P. M. (2019). Being There: International Negotiations as Study Sites in Global Environmental Politics. *Global Environmental Politics*, *19*(2), 4–13. https://doi.org/10.1162/glep_a_00505.

O'Reilly, K. (2012). Ethnographic Returning, Qualitative Longitudinal Research and the Reflexive Analysis of Social Practice. *The Sociological Review*, *60*(3), 518–536. https://doi.org/10.1111/j.1467-954X.2012.02097.x.

Paulsen, K. E. (2009). Ethnography of the Ephemeral: Studying Temporary Scenes through Individual and Collective Approaches. *Social Identities*, *15* (4), 509–524. https://doi.org/10.1080/13504630903043865.

Peña, P. (2010). NTFP and REDD at the Fourth World Conservation Congress: What Is in and What Is Not. *Conservation and Society*, *8*(4), 292–297.

Pink, S., Horst, H. A., Postill, J., et al. (Eds.). (2016). *Digital Ethnography: Principles and Practice*. SAGE.

Platt, J. (1976). *Realities of Social Research*. Sussex University Press.

Pommerolle, M.-E. & Siméant, J. (2011). African Voices and Activists at the WSF in Nairobi: The Uncertain Ways of Transnational African Activism. In J. Smith, E. Reese, S. Byrd & E. Smythe (Eds.), *Handbook on World Social Forum Activism* (pp. 227–247). Routledge.

Postill, J. & Pink, S. (2012). Social Media Ethnography: The Digital Researcher in a Messy Web. *Media International Australia, 145*(1), 123–134. https://doi.org/10.1177/1329878X1214500114.

Rabinow, P. (2003). *Anthropos Today: Reflections on Modern Equipment.* Princeton University Press.

Ransan-Cooper, H., A. Ercan, S. & Duus, S. (2018). When Anger Meets Joy: How Emotions Mobilise and Sustain the Anti-coal Seam Gas Movement in Regional Australia. *Social Movement Studies, 17*(6), 635–657.

Rappaport, J. (2008). Beyond Participant Observation: Collaborative Ethnography as Theoretical Innovation. *Collaborative Anthropologies, 1*(1), 1–31. https://doi.org/10.1353/cla.0.0014.

Reychler, L. (2015). *Time for Peace: The Essential Role of Time in Conflict and Peace Processes.* University of Queensland Press.

Roche, M. (2000). *Mega-events and Modernity.* Routledge. https://doi.org/10.4324/9780203443941.

Rogers, R. (2009). *The End of the Virtual* [Inaugural lecture No. 339]. Amsterdam University Press.

Rucht, D. (2011). Social Forums as Public Stage and Infrastructure of Global Justice Movements. In J. Smith, E. Reese, S. Byrd & E. Smythe (Eds.), *Handbook on World Social Forum Activism* (pp. 11–28). Paradigm Publishers Boulder.

Saerbeck, B., Well, M., Jörgens, H., Goritz, A. & Kolleck, N. (2020). Brokering Climate Action: The UNFCCC Secretariat between Parties and Nonparty Stakeholders. *Global Environmental Politics, 20*(2), 105–127. https://doi.org/10.1162/glep_a_00556.

Salgado, R. S. (2018). The Advocacy of Feelings: Emotions in EU-based Civil Society Organizations' Strategies. *Politics and Governance, 6*(4), 103–114.

Schäfer, M. S., Ivanova, A. & Schmidt, A. (2014). What Drives Media Attention for Climate Change? Explaining Issue Attention in Australian, German and Indian Print Media from 1996 to 2010. *International Communication Gazette, 76*(2), 152–176. https://doi.org/10.1177/1748048513504169.

Schatz, E. (Ed.). (2009). *Political Ethnography: What Immersion Contributes to the Study of Power.* The University of Chicago Press.

Schenuit, F. (2023). Staging Science: Dramaturgical Politics of the IPCC's Special Report on 1.5 °C. *Environmental Science & Policy, 139*, 166–176. https://doi.org/10.1016/j.envsci.2022.10.014.

Schimmelfennig, F. (2002). Goffman Meets IR: Dramaturgical Action in International Community. *International Review of Sociology, 12*(3), 417–437. https://doi.org/10.1080/0390670022000041411.

Schnegg, M. (2021). Ontologies of Climate Change. *American Ethnologist, 48*(3), 260–273. https://doi.org/10.1111/amet.13028.

Schnegg, M. (2023). Becoming a Debtor to Eat: The Transformation of Food Sharing in Namibia. *Ethnos, 88*(2), 392–412. https://doi.org/10.1080/00141844.2021.1887913.

Schuck-Zöller, S., Brinkmann, C. & Rödder, S. (2018). Integrating Research and Practice in Emerging Climate Services — Lessons from Other Transdisciplinary Dialogues. In S. Serrao-Neumann, A. Coudrain, & L. Coulter, L. (Eds.), *Communicating Climate Change Information for Decision-Making* (pp. 105–118). Springer.

Schüssler, E., Rüling, C.-C. & Wittneben, B. B. F. (2014). On Melting Summits: The Limitations of Field-Configuring Events as Catalysts of Change in Transnational Climate Policy. *Academy of Management Journal, 57*(1), 140–171. https://doi.org/10.5465/amj.2011.0812.

Scott, D., Hitchner, S., Maclin, E. M. & Dammert B., J. L. (2014). Fuel for the Fire: Biofuels and the Problem of Translation at the Tenth Conference of the Parties to the Convention on Biological Diversity. *Global Environmental Politics, 14*(3), 84–101. https://doi.org/10.1162/GLEP_a_00240.

Seyfang, G. (2003). Environmental Mega-conferences — From Stockholm to Johannesburg and beyond. *Global Environmental Change, 13*(3), 223–228. https://doi.org/10.1016/S0959-3780(03)00006-2.

Silver, J. J., Gray, N. J., Campbell, L. M., Fairbanks, L. W. & Gruby, R. L. (2015). Blue Economy and Competing Discourses in International Oceans Governance. *The Journal of Environment & Development, 24*(2), 135–160. https://doi.org/10.1177/1070496515580797.

Siméant, J., Pommerolle, M.-E. & Sommier, I. (2015a). Introduction. In J. Siméant, M.-E. Pommerolle & I. Sommier (Eds.), *Observing Protest from a Place* (pp. 11–20). Amsterdam University Press. www.jstor.org/stable/j.ctt16vj27n.4.

Siméant, J., Pommerolle, M.-E. & Sommier, I. (Eds.). (2015b). *Observing Protest from a Place: The World Social Forum in Dakar*. Amsterdam University Press.

Spittler, G. (2001). Teilnehmende Beobachtung als Dichte Teilnahme. *Zeitschrift Für Ethnologie, 126*(1), 1–25.

Strauss, A., Fagerhaugh, S., Suczek, B. & Wiener, C. (1985). *Social Organization of Medical Work*. University of Chicago Press.

Tarrow, S. G. (2011). *Power in Movement: Social Movements and Contentious Politics*. Cambridge University Press.

Tilly, C. (2006). *Regimes and Repertoires*. University of Chicago Press.

Tilly, C. (2008). *Contentious Performances*. Cambridge University Press.

Urry, J. (2003). Social Networks, Travel and Talk. *The British Journal of Sociology*, *54*(2), 155–175. https://doi.org/10.1080/0007131032000080186.

Vadrot, A. B. M., Langlet, A., Tessnow-von Wysocki, I., et al. (2021). Marine Biodiversity Negotiations during COVID-19: A New Role for Digital Diplomacy? *Global Environmental Politics*, *21*(3), 169–186. https://doi.org/10.1162/glep_a_00605.

Vadrot, A. B. M. & Ruiz Rodríguez, S. C. (2022). Digital Multilateralism in Practice: Extending Critical Policy Ethnography to Digital Negotiation Sites. *International Studies Quarterly*, *66*(3), sqac051. https://doi.org/10.1093/isq/sqac051.

van Asselt, H., Huitema, D. & Jordan, A. (2018). Global Climate Governance after Paris: Setting the Stage for Experimentation? In B. Turnheim, F. Berkhout & P. Kivimaa (Eds.), *Innovating Climate Governance: Moving Beyond Experiments* (pp. 27–46). Cambridge University Press. https://doi.org/10.1017/9781108277679.003.

Van Maanen, J. (2011). *Tales of the Field: On Writing Ethnography*. University of Chicago Press.

van Vree, W. (2011). Meetings: The Frontline of Civilization. *The Sociological Review*, *59*(1), 241–262. https://doi.org/10.1111/j.1467-954X.2011.01987.x.

Vrasti, W. (2008). The Strange Case of Ethnography and International Relations. *Millennium*, *37*(2), 279–301. https://doi.org/10.1177/0305829808097641.

Wall, S. S. (2015). Focused Ethnography: A Methodological Adaptation for Social Research in Emerging Contexts. *Forum Qualitative Sozialforschung / Forum: Qualitative Social Research*, *16*(1), 1–15, Article 1. https://doi.org/10.17169/fqs-16.1.2182.

Watson, C. W. (Ed.). (1999). *Being There: Fieldwork in Anthropology*. Pluto Press. https://doi.org/10.2307/j.ctt18fs9h3.

Welch-Devine, M. & Campbell, L. M. (2010). Sorting Out Roles and Defining Divides: Social Sciences at the World Conservation Congress. *Conservation and Society*, *8*(4), 339–348.

Well, M., Saerbeck, B., Jörgens, H. & Kolleck, N. (2020). Between Mandate and Motivation: Bureaucratic Behavior in Global Climate Governance. *Global Governance: A Review of Multilateralism and International Organizations*, *26*(1), 99–120. https://doi.org/10.1163/19426720-02601006.

Werron, T. (2015). What Do Nation-States Compete For?: A World-Societal Perspective on Competition for 'Soft' Global Goods. In B. Holzer, F. Kastner, & T. Werron (Eds.), *From Globalization to World Society. Neo-Institutional and Systems-Theoretical Perspectives* (pp. 85–106). Routledge. https://pub.uni-bielefeld.de/record/2906249.

Westerståhl, J. (1983). Objective News Reporting: General Premises. *Communication Research*, *10*(3), 403–424. https://doi.org/10.1177/009365083010003007.

Wilkens, J. & Datchoua-Tirvaudey, A. R. (2022). Researching Climate Justice: A Decolonial Approach to Global Climate Governance. *International Affairs*, *98*(1), 125–143.

Wilshusen, P. R. (2019). Environmental Governance in Motion: Practices of Assemblage and the Political Performativity of Economistic Conservation. *World Development*, *124*, 1–14. https://doi.org/10.1016/j.worlddev.2019.104626.

Wolcott, H. F. (1999). *Ethnography: A way of seeing*. Rowman Altamira.

Acknowledgements

The research was funded by the Deutsche Forschungsgemeinschaft (DFG, German Research Foundation) under Germany's Excellence Strategy (EXC 2037 'CLICCS—Climate, Climatic Change, and Society', Project 390683824).

We would like to thank the members of our research team at COP26 in Glasgow, Emilie d'Amico, Alvine Datchoua-Tirvauday, Christopher Pavenstädt, Ella Karnik Hinks, Felix Schenuit and Jan Wilkens, and our colleagues at the Research Seminar on Social Science Climate Research at the University of Hamburg for their friendly support and constructive criticism, as well as Jean Foyer, Michael Schnegg and two anonymous reviewers for helpful comments on earlier versions of this manuscript. Paul Reeve's careful proofreading has greatly improved the Element.

About the Authors

Stefan Cihan Aykut is Mercator Professor of Sociology, in particular Social Dynamics of Ecological Transformation, and Director at the Center for sustainable society research at Universität Hamburg. His research focuses on societal responses to global ecological crises. He was awarded the Heinz Maier-Leibnitz-Prize by the German Research Foundation in 2017.

Max Braun is a research associate at Universität Hamburg where he works in a project funded by the German Research Foundation investigating academic travel and conferences from a sociological perspective. His research currently focuses on science studies, the sociology of interaction, qualitative methods and ethnography.

Simone Rödder is Assistant Professor of Sociology, with a focus on Science Studies, and Principal Investigator in the DFG Cluster of Excellence "Climate, Climatic Change and Society" at Universität Hamburg. Her current research focuses on climate futures, climate movements and the medialisation and politicisation of expertise.

Cambridge Elements

Earth System Governance

Frank Biermann
Utrecht University

Frank Biermann is Research Professor of Global Sustainability Governance with the Copernicus Institute of Sustainable Development, Utrecht University, the Netherlands. He is the founding Chair of the Earth System Governance Project, a global transdisciplinary research network launched in 2009; and Editor-in-Chief of the new peer-reviewed journal *Earth System Governance* (Elsevier). In April 2018, he won a European Research Council Advanced Grant for a research program on the steering effects of the Sustainable Development Goals.

Aarti Gupta
Wageningen University

Aarti Gupta is Professor of Global Environmental Governance at Wageningen University, The Netherlands. She is Lead Faculty and a member of the Scientific Steering Committee of the Earth System Governance (ESG) Project and a Coordinating Lead Author of its 2018 Science and Implementation Plan. She is also principal investigator of the Dutch Research Council-funded TRANSGOV project on the Transformative Potential of Transparency in Climate Governance. She holds a PhD from Yale University in environmental studies.

Michael Mason
London School of Economics and Political Science

Michael Mason is a full professor in the Department of Geography and Environment at the London School of Economics and Political Science. At LSE he is also Director of the Middle East Centre and an Associate of the Grantham Institute on Climate Change and the Environment. Alongside his academic research on environmental politics and governance, he has advised various governments and international organisations on environmental policy issues, including the European Commission, ICRC, NATO, the UK Government (FCDO), and UNDP.

About the Series

Linked with the Earth System Governance Project, this exciting new series will provide concise but authoritative studies of the governance of complex socio-ecological systems, written by world-leading scholars. Highly interdisciplinary in scope, the series will address governance processes and institutions at all levels of decision-making, from local to global, within a planetary perspective that seeks to align current institutions and governance systems with the fundamental 21st Century challenges of global environmental change and earth system transformations.

Elements in this series will present cutting edge scientific research, while also seeking to contribute innovative transformative ideas towards better governance. A key aim of the series is to present policy-relevant research that is of interest to both academics and policy-makers working on earth system governance.

More information about the Earth System Governance project can be found at: www.earthsystemgovernance.org.

Cambridge Elements⁼

Earth System Governance

Elements in the Series

The Carbon Market Challenge: Preventing Abuse Through Effective Governance
Regina Betz, Axel Michaelowa, Paula Castro, Raphaela Kotsch, Michael Mehling, Katharina Michaelowa, and Andrea Baranzini

Addressing the Grand Challenges of Planetary Governance: The Future of the Global Political Order
Oran R. Young

Adaptive Governance to Manage Human Mobility and Natural Resource Stress
Saleem H. Ali, Martin Clifford, Dominic Kniveton, Caroline Zickgraf, and Sonja Ayeb-Karlsson

The Emergence of Geoengineering: How Knowledge Networks Form Governance Objects
Ina Möller

The Normative Foundations of International Climate Adaptation Finance
Romain Weikmans

Just Transitions: Promise and Contestation
Dimitris Stevis

A Green and Just Recovery from COVID-19?: Government Investment in the Energy Transition during the Pandemic
Kyla Tienhaara, Tom Moerenhout, Vanessa Corkal, Joachim Roth, Hannah Ascough, Jessica Herrera Betancur, Samantha Hussman, Jessica Oliver, Kabir Shahani, and Tianna Tischbein

The Politics of Deep Time
Frederic Hanusch

Trade and the Environment: Drivers and Effects of Environmental Provisions in Trade Agreements
Clara Brandi and Jean-Frédéric Morin

Building Capabilities for Earth System Governance
Jochen Prantl, Ana Flávia Barros-Platiau, Cristina Yumie Aoki Inoue, Joana Castro Pereira, Thais Lemos Ribeiro, and Eduardo Viola

Learning for Environmental Governance: Insights for a More Adaptive Future
Andrea K. Gerlak and Tanya Heikkila

Collaborative Ethnography of Global Environmental Governance: Concepts, Methods and Practices
Stefan C. Aykut, Max Braun, and Simone Rödder

A full series listing is available at: www.cambridge.org/EESG

Milton Keynes UK
Ingram Content Group UK Ltd.
UKHW021954050624
443699UK00006B/24

9 781009 476041